THE MOON

KOSMOS

A series exploring our expanding knowledge
of the cosmos through science and technology
and investigating historical, contemporary
and future developments as well as providing
guidance for all those interested in astronomy.

Series Editor: Peter Morris

Already published:

Jupiter William Sheehan and Thomas Hockey
The Moon Bill Leatherbarrow
The Sun Leon Golub and Jay M. Pasachoff

The Moon

Bill Leatherbarrow

REAKTION BOOKS

For Alex and Zoe

Published by Reaktion Books Ltd
Unit 32, Waterside
44–48 Wharf Road
London N1 7UX, UK
www.reaktionbooks.co.uk

First published 2018
Copyright © Bill Leatherbarrow 2018

Printed and bound in China by 1010 Printing International Ltd

A catalogue record for this book is available from the British Library

ISBN 978 1 78023 914 9

CONTENTS

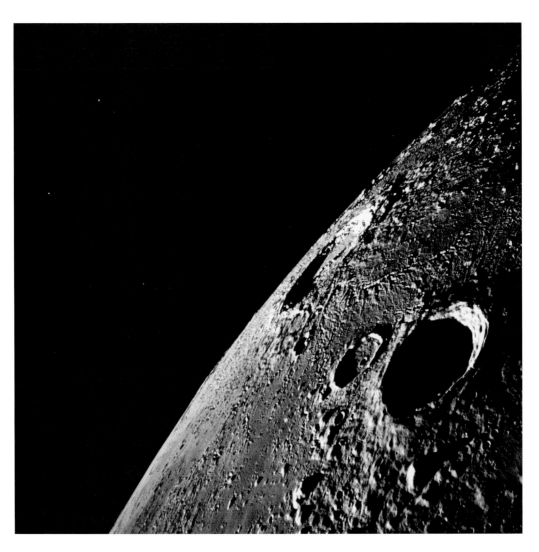

The surface of the Moon.

Preface

The aim in writing this book has been not only to provide an accessible account of current lunar science, but to uncover the history of observation and discovery that gave rise to that science. What we know about the Moon today does not exist in isolation, but has emerged from an unfolding process that has seen awareness of our satellite pass through various stages and revolutions. We cannot fully understand our Moon unless we also understand that process of discovery.

We shall see in the course of the following chapters just how radically our perception of the Moon has shifted over time as our knowledge and techniques have improved. Initially an inexplicable, and at times alarming, presence in our skies – one that gave rise to all kinds of superstitions and religious beliefs – the Moon then evolved into an embodiment of our increasing understanding of celestial mechanics, as we uncovered the mysteries of its movements and appearances. Subsequently, with the invention of the telescope it was revealed as another world, one that was perhaps similar to our own and possibly capable of supporting life. With the advent of the Space Age that dream was laid aside; our Moon turned out to be a rocky, airless and uninhabited world, but one that was no less fascinating. A geologist's paradise, it provides for those willing to decipher it a dramatic account of much of our solar system's history and an important key to the understanding of worlds elsewhere.

That process of discovery is one that can be replicated by the amateur observer, and this book also sets out to provide practical guidance for the owner of a small telescope. It is possible to understand our Moon by reading books about it, but nothing brings home the true meaning of science better than active participation in it.

ONE

OUR COMPANION MOON: FROM MIRROR TO MUSEUM

The Moon has always been the most obvious feature in our night sky. It is our nearest celestial neighbour, at an average distance of 384,400 km, and with an equatorial diameter of 3,476 km it subtends an angle of approximately half a degree in the sky, large enough to display significant detail even to the unaided eye. Moreover, since it describes a regular orbit around the Earth (or, more accurately, around the common centre of gravity of the Earth–Moon system) in a period of only 27.3 days, it is more or less a constant presence in our skies, going through its familiar monthly phases of New Moon, Half-moon and Full Moon. This period of revolution is about two days shorter than the 29.5 days that the Moon takes to complete that cycle of phases (its synodic period, or the time between successive New Moons). The reason for this is that while the Moon is revolving around the Earth, the Earth is in turn continuing along its own orbit around the Sun. So the Moon has to travel that bit farther each month in order to produce the alignment of Sun, Moon and Earth that gives rise to New Moon (or indeed any other phase).

Our Moon has drawn observers since the dawn of humankind, and all have tried to make sense in their own ways of the puzzles it poses and the questions it raises. What is it? How did it get there? What should we make of the appearances, or phases, it presents? Why does it move as it does? What does it mean to humankind, and what opportunities and threats might it pose?

In ancient times it was thought that the Moon was a mirror, its mottled face of dark and light patches nothing more than a reflection of the main oceans and land masses of the Earth. The idea appears to have originated in classical Greece but it migrated subsequently to the folklore of medieval Europe and the Middle East, and for a while it was considered as a possible indirect means of mapping the Earth's topography.[1] Such a view in its literal sense did not long withstand the more rigorous tests of time and observation, but figuratively it has always possessed a certain truth: the Moon has indeed long acted as a reflection of humankind's practical needs, as well as of its desires, hopes, fears and myths. Not only did it serve to illuminate man's nocturnal activities and help banish the dangers and terrors of the night, but it appeared to regulate the tides. What is more, its regular cycle of phases and motions across the sky served to provide our ancestors with one of the earliest means of keeping and measuring time.

Over the centuries the Moon also came to embody the self-image and cultural beliefs of human society. These ranged from the generalized anthropomorphism of the 'man in the Moon', whereby human features were discerned in the Moon's visual aspect as seen with the naked eye, through to the specific Moon legends espoused at some time or other by most pre-scientific civilizations throughout the world.

Many early religions had cults of Moon-worship, and these culminated in the Greek and Roman goddesses Selene and Luna, who were second only to the male Sun-gods Helios and Sol. Both Selene and Luna were believed to exercise an influence upon fertility, a belief encouraged by the similarity in the duration of the lunar synodic period and the female menstrual cycle. Indeed, the word 'menstrual' comes from the Latin word *mensis* (month), which in turn is derived from *mene*, the ancient Greek word for lunar months or 'moons'.

The Moon has also been held accountable for other human conditions, most notably those of madness and psychological disorder. Once again, etymology cements the link through the terms 'lunacy' and 'lunatic'. How that link came into being is uncertain, although in ancient Greece and Rome water was seen as the key: just as the Moon held sway over the tides, so it might control the water that made up the bulk of the human body and brain. The idea is clearly unsound, but it is firmly entrenched in our cultures and myths even to this day. Periods of Full Moon reportedly still give rise to a greater incidence of erratic and irrational behaviour (although this, if true, is more likely to be attributable to the brighter nights during such periods). The link with lunacy is easily extended into one between the Moon and the supernatural, and that is something that has been fully exploited in literature, art and all forms of popular culture, so that it requires no further consideration here.

The catalogue of influences, both benign and malign, that mankind has historically ascribed to our satellite is of course of great intrinsic interest, and it tells us much about human cultural history. However, it contributes little or nothing to our scientific understanding of the Moon itself.[2] Lunar science truly began only when man, instead of projecting his own imaginative cultural, emotional and psychological predilections onto the Moon, began instead to regard the Moon in its own right, as a natural phenomenon to be studied and understood. We might regard that shift as one of learning how to adopt an astronomical understanding of the Moon's appearance and behaviour in our night sky. The etymology of the word 'astronomy' is relevant here, for it is derived from the ancient Greek *astron* (star) and *nomos* (law; arrangement; regulation). In other words, it implies an attempt to understand the laws governing a phenomenon, which is a pretty good general description of what science is!

Astronomy is perhaps the oldest of the sciences – simply because of the easy accessibility of the night sky – and its origins go back to the ancient Babylonians, Chinese, Egyptians, Iranians, Mayans and Indians. But for a truly seismic shift in the development of scientific astronomy, we are indebted mostly to the philosophers of ancient Greece. This is no place to examine their achievement in detail,[3] but we might note in passing the number of Greek philosophers whose significance in the history of astronomy has been recognized by the allocation of their names to the most prominent lunar craters.

These include the polymath Ptolemy, whose work provided one of the earliest efforts to draw up a comprehensive cosmological system. Ptolemy attempted to measure the size of the known universe and to explain its phenomena in terms of a celestial mechanism that had Earth at its centre, around which everything else revolved in a series of circular orbits. He was not the first to advocate a geocentric universe, for such a notion fitted well into contemporary notions of Earth's (and humankind's) primacy in creation; and indeed, those who later dissented from such a view would feel the full force of reprisal, especially from the custodians of religious orthodoxy. But Ptolemy did refine geocentric cosmology to the extent that it would survive more or less intact until challenged by the Copernican Revolution of the sixteenth century. That revolution displaced the Earth from its position of primary importance at the centre of creation and proposed instead a heliocentric model that, once refined by the later work of Johannes Kepler and Isaac Newton on the laws of planetary motion and gravitation respectively, would provide a sound basis for the development of modern cosmology.

Early astronomical science in the pre-telescopic age was thus dominated by an observational approach that was centred on celestial mechanics, that is, the study of the motions of the heavenly bodies and their interrelationships. In the Ptolemaic system the Moon, like everything else, revolved around the Earth, although the forces determining that relationship were not to be properly understood until Newton developed his theory of gravitation in the latter part of the seventeenth century. This recognition marked a major shift in humankind's understanding of the Moon: no longer was it merely a mirror, passively reflecting subjective human expectations. Now it was an objective astronomical entity in its own right, obeying the laws of celestial mechanics as they were understood at the time and playing its own role in the grand ballet of the universe as it regularly described its unique set of movements and appearances across the night sky. One of the problems with the geocentric system was its insistence upon circularity of the orbits followed by the Moon and the planets. If the Moon's path around the Earth were truly circular (and how could it be otherwise in a perfect, divinely created universe, for the circle was regarded by the Greeks as a perfect form), then the Moon would always be at the same distance from Earth at all points in its orbit and would thus always appear the same size in our sky. Yet it was apparent even to the earliest observers that such was not the case: the Moon appeared bigger at some times than at others.

This phenomenon was partly created by the well-known (if still imperfectly understood) Moon illusion, whereby the Moon's disc seems larger when it is low above the horizon and may be readily compared with terrestrial objects like buildings and trees. However, the Moon also displayed perceptible changes in size over time even when it was placed high in the sky. In order to address this and other anomalies in the motion of the Moon, Ptolemy introduced the concept of epicycles. This was a truly cumbersome idea whereby the Moon did not circle the Earth directly, but instead revolved in a small circle about a point, or deferent, which itself described a larger orbit around the Earth. The introduction of epicycles also allowed Ptolemy to explain the occasional retrograde motions described by some of the planets, most notably Mars and Jupiter.

However, the system pushed credulity to the limit. It was only in 1609, after the acceptance of the heliocentric model, that Kepler was able to account for the apparent changes in the Moon's size by recognizing that the orbits of the Moon and the planets were not circular, but elliptical. As a result, the Moon appeared smaller when more distantly situated at the extreme of the major axis of its path around the Earth. The period of greatest distance, when the centre of the Moon can be as far away from Earth's centre as 406,697 km, is known as apogee. However, at its closest point, or perigee, the distance is only 356,410 km.[4]

For the Moon's first observers, another puzzling aspect of its behaviour was its phases, or apparent changes of shape, that it underwent in the course of each month. Primitive superstitions had no real explanation for this phenomenon, but celestial mechanics did and early lunar science quickly recognized that the Moon's phases were the result of the interrelationship of three bodies: the Earth, the Moon and the Sun. The Moon shines by reflecting the light of the Sun (although it is a poor reflector with an average albedo, or reflectivity, of less than 10 per cent). It is also a spherical body, which means that at any one moment only 50 per cent of its surface is illuminated. As it follows its orbit around the Earth, different amounts of the illuminated half are presented towards the planet. Thus it appears to grow from a waxing crescent, through half-phase (First Quarter), to fully illuminated, and then to wane back again through half (Last Quarter) and crescent phases. At new, the Moon is approximately between us and the Sun, so that its unilluminated hemisphere is turned towards us and we do not see it. At full, the Moon is opposite the Sun in our sky and we see the entirety of its illuminated disc. In between, varying amounts of the sunlit side are seen, giving rise to the crescent, half and gibbous phases. In the representation below, the light of the Sun is coming from the right.

A further obvious aspect of the Moon's appearance, and one that would have been apparent even to its earliest observers, is that it always keeps the same face turned towards us – a face characterized by the familiar features of the 'man in the Moon'. Before Newton's theory of gravitation, this would have been difficult to explain, but we now know that it is because over time the Earth's gravitational

First Quarter

Waxing Gibbous

Waxing Crescent

Full

New

Waning Gibbous

Waning Crescent

Third Quarter

Phases of the Moon.

field has slowed down the Moon's rotation so that it is now tidally locked or 'captured'. Consequently it turns once on its axis in roughly the same time that it takes to complete one revolution around the Earth.

In fact, over a period of time we can see more than 50 per cent of the Moon's surface because of the effects of libration, the apparent rocking of the Moon on its axis. Libration in longitude, which permits us to peer around the east and west edges of the

15

Moon, is a result of the Moon's elliptical, rather than circular, path around the Earth. This causes variations in the speed at which the Moon orbits: it moves more quickly through the part of its elliptical path that is closer to the Earth and more slowly elsewhere. However, its axial rotation period remains constant, and this allows the Moon to appear to rock from side to side, exposing the east and west libration zones in turn. There is also libration in latitude, caused by the Moon's axial tilt of about 6.7° relative to its orbit. This allows us glimpses of the north and south libration zones. There are also daily variations (diurnal libration), but the effects of these are small and we do not need to consider them here. As a result of the combined effects of these librations, we can see over time a total of about 59 per cent of the Moon's surface, although features in the zones exposed by libration are significantly foreshortened and difficult to observe.

The ancients were also puzzled by eclipses. From time to time they observed how the Sun was blotted out, either partially or totally, and the Earth plunged into semi-darkness – an eclipse of the Sun. On other, rather more frequent, occasions, the Full Moon would turn a dark red colour, sometimes fading from view almost entirely – an eclipse of the Moon. Early superstitions saw malign significance in those phenomena. For example, both solar and lunar eclipses were often ascribed to the celestial body in question being eaten by a beast (usually a dragon, frog or fox, depending on the myth) or swallowed up by a demon, in which case the offender might be driven away by making loud noises. Once again, increased understanding of celestial mechanics, and especially of the interrelationship of the Sun, Earth and Moon, eventually afforded a more scientific explanation.

Because of the inclination of its orbit around the Earth (approximately 5°), the Moon at its new and full phases normally passes slightly north or south of the plane of the Earth's orbit around the Sun. On the relatively rare occasions when it does not, we see an eclipse. A total solar eclipse occurs when the New Moon passes directly in front of the Sun, and the phenomenon is dependent upon a most remarkable coincidence: the Sun is about four hundred times larger than the Moon, but it is also about four hundred times farther away, so that the two bodies appear almost exactly the same size in the sky as seen from Earth. If the Moon were smaller or more distant it would not appear large enough to cover the Sun's disc entirely and we would see only annular eclipses, where a ring of the bright solar disc remains visible around the Moon. If it were closer or larger it would obscure the Sun's beautiful corona, or outer atmosphere, which is the most spectacular feature of a total eclipse. We now know that because of tidal effects the Moon is slowly receding from the Earth, and a time will come in the distant future when the great spectacle of a total solar eclipse will no longer be seen. However, as the rate of recession is only about 4 cm per year, we need not be overly anxious just yet!

A total lunar eclipse, on the other hand, occurs when the Full Moon passes through the central part, or umbra, of the Earth's shadow. The conditions here are less critical, for the Moon is much smaller than the shadow and takes some time to pass through it. This also means that eclipses of the Moon occur much more frequently than those of the Sun at any given location on the Earth's surface. Both solar and lunar eclipses may be partial when the alignment of the three bodies is not perfect enough to produce totality, but such events are far less spectacular than total eclipses. Moreover, on occasions the Moon might avoid the central umbra of Earth's shadow but still pass through the partial shadow, or penumbra, in which case the drop in brightness is hardly discernible.

With increased scientific understanding came more sophisticated attempts to understand not just the motions of the Moon but also its nature and origin. Crude superstition and mythology were gradually replaced by hypotheses based upon current physical

17

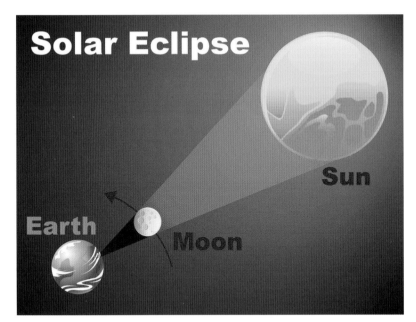

Positions of the Sun, Moon and Earth during a total solar eclipse.

understanding, even if that understanding subsequently turned out to be incomplete or even wrong. Uncertainty still attends our knowledge of how the Moon came into being, but over time several theories have been seriously considered before being discarded in the light of growing evidence. These included the belief that the Moon and the Earth condensed together as a double planet from the original solar nebula, a disc composed of primeval gases and dust drawn together by mutual gravitational attraction which eventually condensed to form the Sun, planets and other bodies making up our solar system. If true, such a theory would require both the Moon and Earth to be of similar age and composition, something that could not be tested until samples of lunar rock were returned by the Apollo missions of the 1960s and '70s. As it turned out, the Earth and Moon did prove to be of similar age, but there were enough differences in composition to cast doubt on the nebular hypothesis. In particular, the Moon appears to be significantly less dense than the Earth, a point we shall return to shortly.

A further theory, advanced in the latter part of the nineteenth century by George Darwin (son of Charles) and the geologist Osmond Fisher, proposed that the Moon had once been part of a rapidly rotating Earth, but had been flung off by centrifugal force, leaving the basin that is now filled by the Pacific Ocean. There were many things wrong with this idea, not least the incongruities of size between the Moon and the hole supposedly left on Earth. Moreover, this theory also fails to account for the different Earth–Moon densities later revealed by Apollo.

Two other theories suggested an extraneous origin for our Moon: the first, supported by the twentieth-century American chemist Harold C. Urey, argued that the Moon had formed elsewhere and had migrated close enough to Earth to be captured by its gravitational field. This was by no means a silly idea, as we now recognize examples elsewhere in the solar system of smaller satellites being captured by larger bodies. But such an outcome requires a precise and unlikely set of circumstances: too direct an

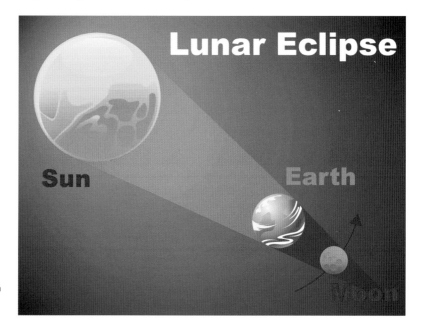

Positions of the Sun, Earth and Moon during a total lunar eclipse.

approach and the Moon would have impacted the Earth, whereas too tangential a pass would have resulted in a fly-by, rather than capture. The other 'extraneous' theory proposed that the Moon formed early in the history of the solar system out of debris from colliding planetesimals that coalesced in the Earth's vicinity.

The greater understanding of the Moon's composition provided by laboratory analysis of returned rock samples allowed two American astronomers, William K. Hartmann and Donald R. Davies, to resurrect in 1975 an idea originally advanced in 1946 by the Canadian geologist Reginald A. Daly. The 'giant impact hypothesis', as it subsequently became known, is not unproblematic, but it is probably the best explanation of the Moon's origin that we have to date. The early solar system was a dangerous and violent place, and Hartmann and Davies modelled a sequence of events occurring between 4 and 4.4 billion years ago that started with the proto-Earth, still in a molten state, being struck at an oblique angle by another protoplanet about the size of Mars. This hypothetical body has acquired the name Theia. Theia was largely absorbed by Earth on impact, but the collision saw an amount of the lighter mantle material that had differentiated and risen to the surface of both bodies thrown off and into orbit around the surviving Earth. This material subsequently accreted to form the Moon.

The hypothesis leaves many questions unanswered, but it does at least account for the Moon's relatively small core and overall lower density. It also helps to explain another unusual feature of the Earth–Moon system: the large size of our satellite relative to its parent planet. Although the Moon is significantly smaller and less massive than Earth, to the extent that the common centre of gravity, or barycentre, of the Earth–Moon system lies some 1,700 km *below* the Earth's surface, most other moons in the solar system are much smaller still relative to their parent body. (The one exception is Charon in relation to the dwarf planet Pluto.) Indeed, it might make more sense to consider the Earth–Moon system as a double-planet, rather than as a hierarchical system in which the Moon plays the lesser role of a mere attendant (the original meaning of the word 'satellite'). The Moon is our companion, rather than our attendant, for we now know that it probably played a crucial role in the emergence and development of Earth as a planet capable of sustaining life.

There are many hypotheses that have been advanced in support of this contention. First, it has been argued that the Theia collision would have contributed additional iron to the core of the proto-Earth and the momentum imparted by the glancing impact would have set the Earth rapidly rotating. That combination would have given rise to the Earth's strong magnetic field, which has since served to deflect dangerous particles in the solar wind that would have been harmful to life had they reached the surface. The same impact might also have contributed to Earth's plate tectonics, which have served over time to recycle the carbon dioxide produced by volcanic activity and prevent the sort of runaway greenhouse effect we see on our neighbour Venus.

Then, once the Moon had coalesced from the material thrown off from the Theia impact, its gravitational influence would have served over time both to slow the Earth's original rapid rotation and to stabilize its axial tilt within fixed limits. This created our benign cycle of seasons and avoided the catastrophic axial shifts that we suspect have occurred on Mars in the course of that planet's history and which might have militated against the emergence or survival of life on that world. Axial instability has also created extreme seasonal changes elsewhere in the solar system, such as those we see on Uranus, where the axial tilt of 98° means that different parts of the planet are presented towards the Sun for lengthy periods and each pole endures a night lasting for 21 Earth years!

Finally, the tides raised on Earth by the Moon have served to enrich our oceans with the sort of minerals that are essential for the emergence of marine (and subsequently terrestrial) life.[5]

Following the Theia collision the Earth–Moon system, like other bodies in the young and still unstable solar system, was to suffer further catastrophic bombardment by planetesimals, comets and asteroids that were flung around in a devastating game of celestial billiards created by the orbital instabilities and gravitational influence of Jupiter and the other large outer planets. Much of this occurred some 4.1–3.8 billion years ago during a period known as the Late Heavy Bombardment, and we shall have much more to say about it later when we come to consider the origins of the Moon's surface features.

The Earth's later evolution, and especially the role played in its development by atmosphere, water, weather, wind and plate tectonics, led to evidence of that bombardment being largely erased from the surface of our home planet. But the Moon's smaller size and lesser density meant that over time it was unable to retain any significant atmosphere it might once have had and it cooled too quickly to experience the regular resurfacing caused by plate tectonics or enduring volcanic activity. The Moon did have a volcanic past, of which more later, but that was largely finished some 2.5 billion years ago (although smaller-scale volcanic activity continued beyond that time). The result was that, compared to Earth, the Moon experienced little in the way of sustained surface erosion, apart from the 'space weathering' caused by the action of the solar wind and by micrometeoroidal impacts. It therefore still displays the stark and dramatic scars of past impacts, and it is a veritable wonderland for the telescopic observer who wishes to explore the intricacies of its topography.

Thanks to ongoing exploration by robotic spacecraft, we continue to learn more and our relationship with our companion Moon continues to evolve. Once merely a mirror held up solely to confirm humankind's superstitions, it is now a gloriously informative museum revealing to us the true nature of our solar system's violent past. In the next chapter we shall see how the invention of the telescope allowed us to enter that museum and begin a new stage in our journey of discovery.

TWO

THE MOON AS A WORLD: OBSERVATION AND DISCOVERY IN THE TELESCOPIC AGE

The invention of the telescope is popularly attributed to the Dutch spectacle-maker Hans Lippershey of Middelburg, although similar devices were almost certainly to be found elsewhere even before that day in 1608 when, as the story has it, an accidental arrangement of lenses by Lippershey's children was found to enlarge the appearance of a distant church weathervane. What is not in doubt is that Lippershey tried to stake a formal claim to the invention in October of 1608, but he was unsuccessful in establishing a patent and, in the months that followed, news of the device spread throughout Europe as its potential military value became clear.[1]

Among those who recognized that potential was Galileo Galilei (1564–1642), then Professor of Mathematics at the University of Padua. Keen to advance his career, Galileo was at first more interested in the terrestrial applications of the new instrument and he quickly constructed ever more powerful versions with a view to impressing possible patrons among the dignitaries of nearby Venice. But, ambitious as he was, Galileo was also scientifically curious and alert to the telescope's astronomical implications, with the result that on 30 November 1609 he turned his primitive device, magnifying twenty times but with a painfully narrow field of view, to the waxing crescent Moon. Later that year and into the following he went on to point his telescope at Jupiter, observing that planet's four major moons and in the process dealing a severe blow to those who still clung to the idea of a geocentric universe. That blow was reinforced by his later discovery of the phases of Venus, for such an appearance was possible only if Venus circled the Sun in a nearer orbit than Earth's.

In the words of David Whitehouse, the moment when Galileo turned his telescope to the Moon was 'one of those rare moments when the universe changed, when the speculations and prejudices of antiquity fell away'. In the observational drawings that he made on that and subsequent nights, we see for the first time 'the modern moon, a planet stripped of symbolism and myth, a stark world awaiting exploration and discovery'.[2] Before Galileo's eyes the Moon was revealed as more than just a disc in the sky, more than just a cog in the celestial mechanism: it was another world, not identical to ours to be sure, but similar enough to invite compar-isons. As we shall see in due course, this notion that the Earth was not alone, but was simply one planet in a plurality of worlds, would not sit easily in a Christian Europe structured on the geocentric and anthropocentric notions that the Earth was the heart of God's creation and humankind its culmination.

For Galileo the most immediately striking aspect of the Moon's telescopic appearance was the roughness of its surface, a topography scarred not only by familiar features such as mountains, valleys and smooth expanses that might have been either plains or seas, but by a multitude of features less commonly encountered on Earth: the ubiquitous craters of the Moon. He describes how through the telescope the

Moon appears about thirty times larger, its surface about nine hundred times, and its solid mass nearly 27,000 times larger than when it is viewed only with the naked eye; and

consequently any one may know, with the certainty that is
due to the use of our senses, that the Moon certainly does
not possess a smooth and polished surface, but one rough
and uneven, and, just like the face of the Earth itself, is
everywhere full of vast protuberances, deep chasms, and
sinuosities . . . I have been led to that opinion which I have
expressed, namely, that I feel sure that the surface of the Moon
is not perfectly smooth, free from inequalities and exactly
spherical, as a large school of philosophers considers with
regard to the Moon and the other heavenly bodies, but that, on
the contrary, it is full of inequalities, uneven, full of hollows and
protuberances, just like the surface of the Earth itself, which is
varied everywhere by lofty mountains and deep valleys.[3]

The telescopic revelation of lunar topography thus saw off finally
the classical view of the Moon as an ideally smooth celestial body;
instead it emerged as a real world, imperfect, rocky, pockmarked
and comparable to Earth.[4]

In the months that followed Galileo made several drawings of
the Moon as well as writing an account of his telescopic discoveries
entitled *Sidereus nuncius* ('The Starry Messenger', 1610). In the latter
he also considered such matters as earthshine (when at crescent
phase the unilluminated hemisphere of the Moon is faintly visible
because of light reflected by Earth), as well as the Pythagorean notion
that the Moon's brighter areas might be land and the large darker
patches seas, although he never firmly committed to that view:

if any one wishes to revive the old opinion of the Pythagoreans,
that the Moon is another Earth, so to say, the brighter portion
may very fitly represent the surface of the land, and the darker
the expanse of water. Indeed, I have never doubted that if the
sphere of the Earth were seen from a distance, when flooded
with the Sun's rays, that part of the surface which is land would

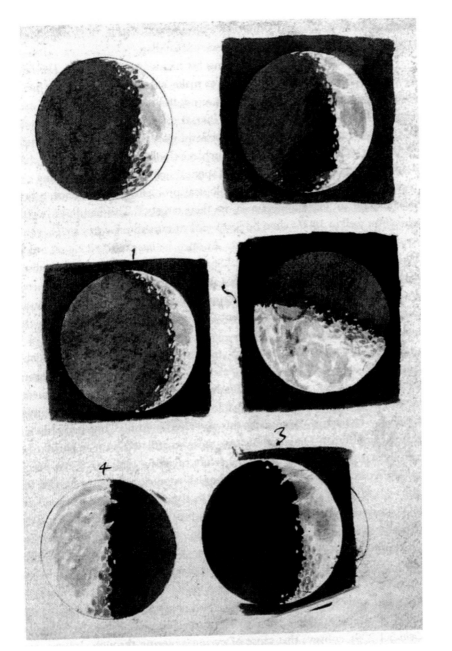

present itself to view as brighter, and that which is water as darker in comparison.[5]

Galileo also recognized the implications of the glancing lighting at the lunar terminator, that stark dividing line between the day and night hemispheres. The dramatic shadows cast at lunar sunrise and sunset not only helped to disclose surface relief and the existence of mountain peaks, but provided a means for determining the height of those peaks by measuring shadow lengths and using simple geometry.

The *Sidereus nuncius* thus makes clear that Galileo's approach to the Moon as seen through the telescope was primarily analytical, rather than cartographic. Indeed, as Allan Chapman points out, Galileo never attempted to make a map of the Moon and his lunar drawings are impressionistic rather than cartographically accurate.[6] John Heilbron takes a similar view when he writes that 'Galileo's renditions move from portraiture almost to caricature.'[7] That is not to undervalue Galileo's observational skills or his abilities as a draughtsman, and it is still perfectly possible to agree with the view expressed by William Sheehan and Thomas Dobbins that 'the quality of Galileo's representations of the lunar surface far surpassed those of his contemporaries' and that they are 'remarkably accurate'.[8] It is just that their 'accuracy' lay in the way they set out to interpret the nature of what he saw, rather than merely conveying the Moon's appearance and the disposition of its surface features.

A more conventionally cartographic approach was that adopted by Galileo's contemporary, the Englishman Thomas Harriot (1560–1621). An accomplished mathematician and astronomer, erstwhile adventurer and explorer, and an acquaintance of Sir Walter Raleigh, Harriot acquired a telescope, or 'dutch truncke' as he called it, early in 1609, and he was the first to use it for observation of the Moon on 26 July that year, some four months before Galileo. His telescope was basic, magnifying only a few times, but as Heilbron

Galileo's drawings of the Moon, 1609.

perceptively writes, 'Harriot saw enough of the moon to map it in the style of a surveyor.'[9] Indeed, he had served as cartographer and surveyor during an expedition to Virginia in 1585–6. Harriot's lunar observations consisted only of rough sketches at first and he did not publish his results until after Galileo had seized the limelight, but by 1610 he had constructed a telescopic map of the lunar surface that depicted the major 'seas', craters, mountain ranges and even the bright ray systems in a way that the telescopic explorer would still recognize today.

Harriot's lunar sketches have occasionally been dismissed as rudimentary and unpolished, and they have been compared unfavourably with those of Galileo;[10] but that is unfair, for the two men were going about their work in entirely different ways. Unlike Galileo, Harriot was indeed 'surveying' the Moon and not trying to convey the drama of its landscape or interpret its topography. He was in effect the first lunar explorer–mapper (or selenographer), adopting an approach that prioritized the accurate charting of the lunar surface while giving less emphasis to the analysis of what he saw. Galileo, on the other hand, might with some justification be described as the first practitioner of a geological approach to the Moon (or selenologist), someone who directed a powerful scientific imagination to the understanding of the surface structures disclosed by telescopic observation. We shall see in the course of this study how those two approaches, both established in that momentous year of 1609, came to inform the subsequent development of lunar science, each pursued often to the exclusion of the other and the two only coming together in a truly easy alliance with the advent of the Space Age.

The cartographic impulse imparted by Harriot to telescopic study of the Moon was continued by many others. This is hardly surprising, for maps were of considerable significance to Europeans in a great age of exploration, annexation and discovery. The burgeoning recognition that the Moon was 'another world' meant

Thomas Harriot's
map of the Moon, 1610.

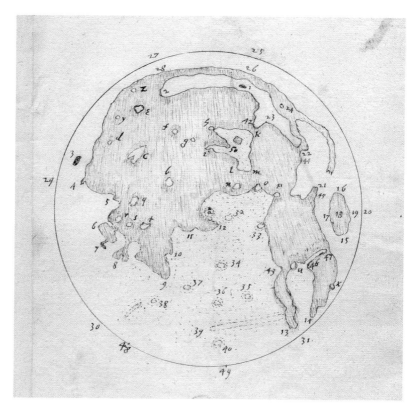

that it, too, was terra incognita awaiting its explorers and mappers.
Cartography and cartographic nomenclature were also ways of
symbolically taming and appropriating newly discovered worlds. It
is thus significant that the first man to produce a proper map of the
Moon based upon careful measurement and with a full complement
of named features was the son of a family of Flemish cartographers
and globe-makers. Michel van Langren (1600–1675), better known
as Langrenus, was a mathematician and engineer, rather than an
astronomer, and his determination to map the Moon accurately
was stimulated by earthly, rather than celestial, considerations.
In particular, the problem of determining longitude at sea to an
accuracy that would allow safe navigation was clearly a priority for
European seafaring nations bent on global exploration. Indeed in

1598 Philip III of Spain had offered a reward of 6,000 ducats for the first to come up with a practical solution.

The principle was simple: longitude could be determined by comparing local time with a standard determination of time from a known location. The problem was that no clock existed that could maintain the required accuracy at sea, and that was to remain the case until John Harrison's invention of the marine chronometer in the late eighteenth century. Langrenus, like many before him, understood that the heavens offered several potential mechanisms for the measurement of time. He favoured the careful timing of the lunar terminator as it uncovered and covered discrete topographical points on the lunar surface. These timings could then be compared to those given in standard ephemerides, which in turn would allow the calculation of local longitude. However, there was no suitably accurate lunar map, and it was this gap that Langrenus set out to fill. The result of his efforts was the publication in 1645 of a 34-cm-diameter engraved map of the Moon's mean visible hemisphere, based on thirty accurate drawings made at various phases and standardized with north up.[11]

Langrenus's map, the Plenilunii lumina austriaca philippica, was the first to address systematically the question of names for the large number of features it showed, suggesting 325 names that had the effect of turning the Moon into 'a place of patronage and homage to powerful men'.[12] The places on his map carry (in Langrenus's own words) 'the proper names of Kings and Princes (who reign today in Europe and are patrons, protectors and promoters of the Sciences and Mathematics), and with the names of other people old and recent who excel in this knowledge and call down praise and fame on themselves'.[13] This approach was not new – after all, Galileo had originally termed the Jovian satellites the 'Medici stars' in honour of his pupil and patron Cosimo, the Grand Duke of Tuscany – but political fortunes change, and neither Galileo's term nor Langrenus's system of lunar nomenclature survived for long.

Langrenus's lunar map, 1645.

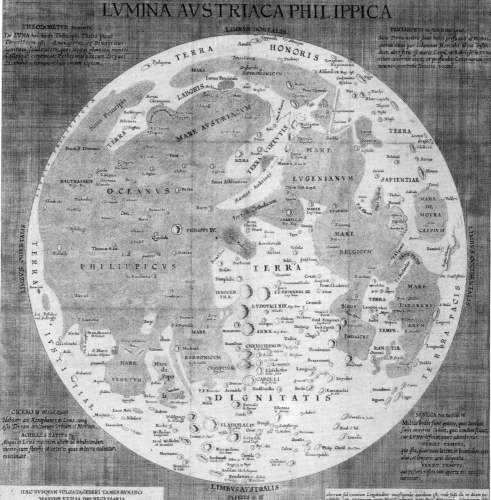

PLENILVNII
LVMINA AVSTRIACA PHILIPPICA

Map of the full Moon with Latin place-names and marginal texts.

Today only three of Langrenus's names are still attached to the same features, including, fittingly, the beautiful crater on the edge of the Mare Fecunditatis that still bears his own name.

The instruments used by the early telescopic observers were unsophisticated, often of poor optical quality and difficult to use. They were of small aperture, which limited their light grasp and the amount of detail they could resolve. They also suffered from a narrow field of view, often not wide enough to encompass the whole of the lunar disc, and were beset by various aberrations as a result of their primitive optical configuration. The most significant of these was sphero-chromatic aberration, where the simple objective lens failed to bring to a common focus both the light from different parts of its aperture and the various spectral colours making up white light. The result was an image marred by false colour and lack of sharpness.

Despite this, real progress was made not just in the mapping of the Moon, but in the scientific understanding of its nature. For example, telescopic scrutiny suggested that the lunar 'seas' might be nothing of the sort, but instead might be waterless plains (although this was only definitively confirmed by Johann Schröter in the eighteenth century). Moreover, even the rudimentary telescopes of the seventeenth century were capable of showing that although the Moon might be another world, it was one that lacked an appreciable atmosphere, since its surface features were always distinctly seen and not obscured by clouds or other atmospheric effects. On 19 March 1637 the English astronomer Jeremiah Horrocks (1618–1641), best known for his observations of the transit of Venus in 1639, watched the Moon occult, or pass in front of, the Pleiades star cluster in the constellation of Taurus. He observed that as the Moon's disc reached each star, the star winked out instantaneously, rather than fading gradually. He realized that this meant two things: first, that from our vantage point on Earth the stars were points of light rather than perceptible

Crater Langrenus.

discs, and they must therefore be unimaginably remote; second, the absence of a fade meant that the light from the star was not extinguished gradually as it would be if it had to pass through a perceptible lunar atmosphere.

Horrocks's deduction is fully confirmed by what we have subsequently learned at first hand: that the Moon is indeed to all intents and purposes an airless world, with an atmospheric density of only 10^{-14} that of the Earth at sea level.[14] It was a view shared also by many subsequent telescopic observers, including such luminaries as Christian Huygens and Tobias Mayer, and was one that had clear implications for the possibility of life on the Moon. For what form, if any, could life take on a world devoid of water and air? However, as we shall see in due course, belief in life on other worlds in our solar system, including the Moon, was an engaging one, and it was to survive the challenges posed by telescopic observation, persisting deep into the twentieth century.

If the existence of life on the Moon was to be confirmed or refuted, bigger and better telescopes would be needed. The problem of the small size of the telescopes used by Galileo and his contemporaries (usually with lenses only an inch (2.54 cm) or so in diameter) would be largely overcome by the mid-seventeenth century when advances in the manufacture of optical-quality glass

allowed the production of larger lenses with greater resolving power. Moreover, it was recognized that sphero-chromatic aberrations diminished in lenses of greater focal length, a solution that unfortunately created further problems of its own. Telescopes assumed unwieldy lengths and had to be supported by tenuous systems of poles, ropes and pulleys. They were extremely difficult to use at the high magnifications needed to study the Moon in detail, which inevitably slowed the development of precise lunar cartography. However, despite that, remarkable progress was made. In the city of Danzig, Johannes Hevelius (1611–1687), a brewer by profession, set up a rooftop observatory with a 3.7-m (12-ft) telescope that he used to observe the Moon. In 1647 he published the results of his observations, a book entitled *Selenographia* containing three whole-Moon maps along with engravings of the Moon at forty different phases.

Hevelius's map of the Moon.

Selenographia was a remarkable achievement and, as Ewen Whitaker points out, it was to remain the standard reference work for the next 150 years.[15] Hevelius's maps and drawings were reasonably accurate considering the observational difficulties encountered, but just as significant were his attempts to depict the Moon's libration zones and his identification of what he took to be volcanoes on the lunar surface (the craters now known as Copernicus and Aristarchus). This latter observation clearly implied that the Moon, like the Earth, might be a world of changes. We shall see just how significant that idea was to become in the later development of selenography.

Meanwhile, the Italian priest Giovanni Riccioli (1598–1671) was taking issue with the system of lunar nomenclature developed by Langrenus as well as with the additional names that had been introduced by Hevelius. Riccioli's lunar map, which appeared in his two-volume *Almagestum novum* (1651), contained names for the Moon's features that were not based upon pomp, precedence or patronage, but which set out instead to rationalize nomenclature.

Hevelius's rooftop observatory, Danzig.

He named larger features, such as
the 'seas', after terrestrial weather,
moods and effects: for example,
Mare Imbrium ('Sea of Rains'), Mare
Tranquillitatis ('Sea of Tranquillity')
and Mare Crisium ('Sea of Crises').
For smaller features, he divided
the Moon's visible face into eight
parts, or octants, and within each
allocated names to great astronomers, scientists and philosophers,
usually grouping together cognate

Deslandres. Cassini's bright spot is just below and to the right of the crater centre.

figures in the same octant. As a Catholic priest, Riccioli could not
openly advocate the heliocentric system, but, as Whitaker has
argued, he attached the names of its greatest proponents –
Aristarchus, Copernicus and Kepler – to three brilliant ray craters
dominating the Oceanus Procellarum.[16]

Riccioli's names were enshrined in the maps drawn up by
the great Italian astronomer Giovanni Cassini (1625–1712), who
went on to become director of the Paris Observatory. Cassini's first
(and larger) map of 1679 did not carry topographical names, but his
later map of 1692 did, and it enjoyed a popular success that helped
ensure the endurance of the Riccioli system to the extent that it went
on to form the basis of all subsequent lunar nomenclature.

Cassini's maps did not in themselves advance lunar cartography
by much, for their accuracy did not match their aesthetic effect.
But the series of observations and drawings on which they were
based are of considerable interest. These contain quite detailed
depictions of specific areas, and among them, as Ivano Dal Prete
has recently discovered, is the earliest known sketch of the greatest
sinuous valley on the lunar surface, now termed the Vallis Schröteri.
The sketch was made in June 1685, nearly a year before Huygens
observed it.[17] Yet more significant was Cassini's observation on

21 October 1671 of what he took to be a cloud on the floor of the vast ruined formation now known as Deslandres. Viewed through modern telescopes this 'cloud' turns out to be merely the brilliant ejecta blanket surrounding a small impact crater – it is nothing special. However, Cassini's interpretation of this bright spot was to have considerable resonance with later observers for, like Hevelius's lunar 'volcanoes', it served to insinuate into selenography the notion of changes on the Moon, something that was to play a large, and not always helpful, role in the way that our understanding of the Moon subsequently unfolded.

A more theoretical and 'geological' approach to volcanism on the Moon was adopted by the English astronomer Robert Hooke (1635–1703). Hooke was a skilful telescopic observer who produced one of the first truly accurate drawings of a lunar feature, the large crater Hipparchus, but he is better known for his musings on lunar crater formation. He was at first struck by the similarity between the craters of the Moon and those produced during an experiment in which he dropped lead balls into wet clay. However, Hooke could have had no inkling of what we know today about our solar system's violent early history during the Late Heavy Bombardment, with the result that he was at a loss to imagine the kind of interplanetary projectile that might have produced impact features of such huge size as the craters on the Moon. Accordingly, he abandoned the impact hypothesis in favour of volcanism following a further experiment in which he watched bubbles form in boiling alabaster. When these burst they, too, left ring formations that were strikingly similar to the Moon's craters. The problem with Hooke's bubble hypothesis was again one of scale, for what could produce and sustain bubbles large enough to give rise to the vast formations found on the Moon?

Hooke also flirted with the idea of life on the Moon. He thought he had detected evidence of vegetation in the form of the dark floors found in some craters, and he believed that larger and better telescopes might one day reveal the existence of living beings, but his ideas on the subject had little impact. What did truly ignite in the popular consciousness the idea of life on other worlds, including the Moon, was the publication in 1686 of the treatise *Conversations on the Plurality of Worlds* by the French writer Bernard le Bovier de Fontenelle. Conceived in the form of dialogues between Fontenelle and a beautiful but fictional marchioness, the book, described by Voltaire as 'one of the best works that ever was written',[18] was to resonate like a bell and fix forever the notion of the Moon and planets as 'other worlds'. This was to have profound significance for how such worlds would be studied in future. Starting from an exposition of the Copernican heliocentric system, Fontenelle went on to challenge those 'scrupulous people who may imagine religion is endangered by placing inhabitants any where but on the earth'.[19] If God created other worlds, he argued, He must have done so in order for them to be habitations for life, for otherwise there would be no point: 'Can you believe that after the earth has been thus made to abound with life, the rest of the planets have not a living creature in them?'[20] Therefore, according to Fontenelle, life most probably existed on the Moon and other planets. However, since the Moon had no atmosphere, life there must be very different from life on Earth, assuming forms that our knowledge could not yet envisage:

Remember the earth has been made known to us by little and little. The ancients positively asserted that the torrid and frozen zones were uninhabitable, from the excessive heat of the one, and the cold of the other . . . We may yet know much more of our own world, and then become acquainted with the moon; till that time we must not expect, because our knowledge is progressive: when we understand our own habitation, we may be permitted to study that of our neighbours.[21]

The truly revolutionary impact of *Conversations on the Plurality of Worlds* is to be found in those words, not just in its advocacy of life

Moret, print showing
Fontenelle meditating
on the plurality of worlds,
1791.

Desfontaines del.   *1791*   *Moret Sculp.*

FONTENELLE MÉDITANT SUR LA PLURALITÉ DES MONDES.

on the Moon, for what Fontenelle was implying was a continuum
of exploration that would affect the nature of lunar and planetary
study for centuries to come. The *terrae incognitae* of the other
worlds in our solar system might indeed have been unknown in
Fontenelle's time, but they were surely *knowable* in the fullness of
time. Moreover, they were knowable as part of the same exploratory
process that had already rendered known what was once unknown
about our Earth. Exploration of other worlds was a continuation of
the same scientific venture that had opened up our own. We shall
see in the next section how the nature of telescopic study of the
Moon and planets in the modern age was to be largely determined
by this geographical approach popularized by Fontenelle.

39

The Great Age of Selenography: Mapping and Understanding Alien Landscapes

By the middle of the seventeenth century the great initial flowering of observational astronomy that followed the invention of the telescope had largely run its course, and study of the Moon was about to enter a period that Joseph Ashbrook has described as 'the long night of selenography'.[22] The reasons for this were largely technical: the imperfect and unwieldy telescopes of the day had revealed as much as they could of the surface of our Moon, and further significant advances had to await the development of new and better optical systems. This would come with the adoption by astronomers of reflecting telescopes, following Isaac Newton's prototype constructed in 1668, and the production of achromatic refractors in the late eighteenth century. The former used mirrors instead of lenses in order to produce colour-free images from instruments of manageable focal lengths, while the compound objective of the achromat allowed the construction of refractors with acceptable colour dispersion.

The honourable exceptions to the long night of selenography were the achievements of Newton and the German cartographer Tobias Mayer (1723–1762). Newton's theory of gravitation and laws of motion, as expounded in his *Philosophiae naturalis principia mathematica* of 1687, provided a new template for the understanding of the Moon's motions, while Mayer produced a small lunar map, published posthumously in 1775, which was based on micrometric measurements in order to determine accurate positions and, for the first time, locate surface features on an orthographic grid of latitude and longitude.

The renaissance in selenography, when it eventually came, was largely steered by British and German astronomers, and observers from those two nations were to dominate the field until well into the twentieth century. Friedrich Wilhelm Herschel (1738–1822) had a foot in both camps. Born in Hanover, Germany, he moved

William Herschel.

to England in 1757, anglicized his forenames to Frederick William, and went on to become a Knight of Hanover and one of England's greatest observational astronomers. Best known for his discovery of the planet Uranus in 1781, he was also an indefatigable deep-sky observer, eventually acquiring a reputation as 'the father of stellar astronomy'. However, in the early part of his career Herschel also paid attention to the Moon, a body on which he believed some form of life to be 'a great probability, not to say almost absolute certainty'.[23] In 1776 he made an observation depicting what he thought to be giant trees on the Mare Humorum in the vicinity of the crater Gassendi, and he was convinced that larger telescopes would in due course reveal evidence of intelligent life.

Herschel also argued the case for active volcanism and topographic changes on the lunar surface, and on the night of 19 and 20 April 1787 he reported seeing

> three volcanoes in different places of the dark part of the new moon. Two of them are already nearly extinct, or otherwise in a state of going to break out, which perhaps may be decided next lunation. The third shows an actual eruption of fire, or luminous matter.

The next night, he said,
'the volcano burns with greater
violence than last night' and
the areas adjacent to it 'seemed
to be faintly illuminated by the
eruption'.[24] Herschel also claimed
to have seen lunar volcanoes even
earlier, in 1780 and 1783, but it
is almost certain that what he
saw in 1787 was nothing more
than the three bright ray craters
Aristarchus, Copernicus and
Kepler faintly illuminated by the
light of earthshine. A prosaic
explanation, but the impact of
Herschel's volcanoes was to
resonate dramatically down future
generations of lunar observers.

One man whose observing
career was inspired by the
discoveries of Herschel was the
German amateur Johann Hieronymus Schröter (1745–1816),
the chief magistrate of the small town of Lilienthal in northern
Germany. Intrigued by Herschel's volcanoes and keen to establish
whether atmospheric or surface changes really occurred on the
Moon, Schröter determined to examine lunar features under varying
conditions of illumination, making detailed drawings of specific
features using telescopes with apertures of between 12 cm (4.75 in.)
and 48.9 cm (19.25 in.). By 1791 he was in a position to publish the
first substantial volume (about 680 pages) of his *Selenotopographische
Fragmente*, a further volume of which appeared in 1802. He would no
doubt have produced more had much of his work not been destroyed
and his observatory sacked during the retreat of Napoleon's forces

Johann Hieronymus
Schröter.

42

in April 1813. As well as charts of specific areas, Schröter's work included the measurement of mountain heights and the monitoring of features such as Alhazen and the dark-floored crater Hevelius, where he believed real topographic changes to have occurred. He also suspected the existence of a tenuous lunar atmosphere, evidence for which he thought he had detected in cusp extensions and attenuations during the crescent phases, appearances that he took to be signs of atmospherically induced twilight. These were almost certainly just peaks and other elevated areas catching the early or late sunlight. More substantially, Schröter finally laid to rest any remaining suspicion that the Moon's dark expanses might be seas of liquid water, for he charted mountain peaks, valley-like rilles, wrinkle ridges and other landforms on their surfaces. He was also the first to apply systematically the term 'crater' to the lunar ring formations.[25]

However, the real significance of Schröter for the later development of selenography is suggested in the title he chose for his major publication. His pursuit of selenographical *fragments* marked a new departure in the history of lunar observation. Until then, telescopic observers had worked towards creating maps of the entirety of the Moon's visible hemisphere. Although Schröter initially intended to produce a 1.2-m (4-ft) map of the lunar surface, he soon scaled down his ambitions and adopted an approach centred upon the detailed charting of selected areas. This retreat from a holistic view of our Moon in favour of one that emphasized instead the significance of close-up views of greater detail was to have considerable impact on those who followed him. Indeed it would largely determine the course of lunar observational science until the advent of the Space Age.

It is clear from the foregoing that by the start of the nineteenth century lunar science in Europe, and especially in Germany and Britain, had become defined by three major themes that were to inform nearly all further exploration during the great age of telescopic astronomy that was to come. Those themes were, first, emphasis on a primarily cartographic approach to practical lunar study; second, an enduring romantic belief in the probability of changes on the Moon (and even lunar life); and third, a general preference for volcanic, rather than impact, theories of lunar surface formation. The cartographic tradition established by Harriot and followed by Langrenus, Hevelius, Riccioli, Cassini, Mayer and others reached its fullest flowering in the nineteenth century, largely due to the efforts of a handful of outstanding German selenographers equipped with determination, a clear sense of purpose and modern, effective telescopes.

The first truly modern map, in the sense that it established many of the conventions followed by later selenographers, was that drawn by Wilhelm Lohrmann (1796–1840), a professional surveyor and cartographer from Dresden. Lohrmann planned a map of around 96 cm (38 in.) in diameter, based on observations made with a 12-cm (4.8-in.) Fraunhofer refractor. The map was to be in 25 sections, a scheme adopted by several subsequent mappers, including members of the later British school, such as Walter Goodacre and H. P. Wilkins. Only four of those sections were published in Lohrmann's lifetime, in 1824, since failing eyesight interrupted his observations and he died early of typhoid fever. He left enough draft work, however, to allow his countryman J.F.J. Schmidt to complete and publish the remaining sections in 1878. Lohrmann's map made artistic use of hachuring and stippling techniques in order to depict surface slopes and albedo differences. It was also remarkably accurate, plotting more than seven thousand craters, including much new detail, on an orthographic grid with lines of latitude and longitude placed at intervals of 5°.[26]

However, by the time Schmidt completed Lohrmann's map, it had already been superseded by the work of two of his contemporaries. 'The most important book written about the moon in the nineteenth century',[27] *Der Mond* by Wilhelm Beer (1797–1850)

Detail from Lohrmann's map of the Moon, from his *Topographie der sichtbaren Mondoberfläche* (Topography of the Lunar Surface), 1824.

and Johann Heinrich Mädler (1794–1874) appeared in 1837. As well as containing their own 96-cm (38-in.) map in four quadrants, based upon observations largely made by Mädler using Beer's 9.5-cm (3.75-in.) refractor, the book summarized what was then known about the Moon and the history of selenography. It also gave detailed measurements and coordinates, and provided a descriptive

45

account of the Moon's surface features – all in a monumental
volume of some four hundred pages. It brought a solid reassurance
to lunar study, along with a sober reassessment of the Moon as a
world, for Mädler's observations had led him to the conclusion that
it was a barren and airless planet – unchanged, unchanging and
devoid of life. The impact of the work was immense, but ambiguous.
No one could afford to overlook *Der Mond*, for the account it gave of
our Moon appeared complete, even if the conclusions it drew did
little to encourage further interest. As Whitaker writes, 'everything
that could be learned about the Moon was there for anyone to read,
and everything that could be observed with a telescope of moderate
size was already mapped.'[28] Further research could surely only tinker
around the edges of Beer and Mädler's remarkable achievement.

Someone who did more than just tinker was the aforementioned
J. F. Julius Schmidt (1825–1884), an enthusiastic observer from
Eutin in northern Germany who rose to become director of the
Athens Observatory in 1858. Schmidt devoted himself assiduously
to lunar observation for more than thirty years, keeping alive the
flame during the years of inactivity following the appearance of
Beer and Mädler's work. As well as completing Lohrmann's map,
he published his own atlas of the Moon in 1878, which included
a map in 25 sections and over 2 m (6 ft) in diameter on which he
plotted the positions of nearly 33,000 craters – 'the finest that was
ever compiled without the aid of photographs.'[29]

But Schmidt is best remembered for an observation he made
of the small crater Linné on 16 October 1866. Located in a relatively
uncluttered part of the Moon, on the surface of the Mare Serenitatis,
Linné had been recorded by several early observers as an ordinary
crater some 8–10 km (5–6 miles) in diameter and quite deep.
However, Schmidt now saw it as a small whitish spot and he
concluded that a real topographic change had occurred, the original
Linné having been obliterated by some form of eruptive process.
Others, including Mädler, were sceptical, and indeed the evidence

Beer and Mädler's map, northeast quadrant, from their *Der Mond* (1837).

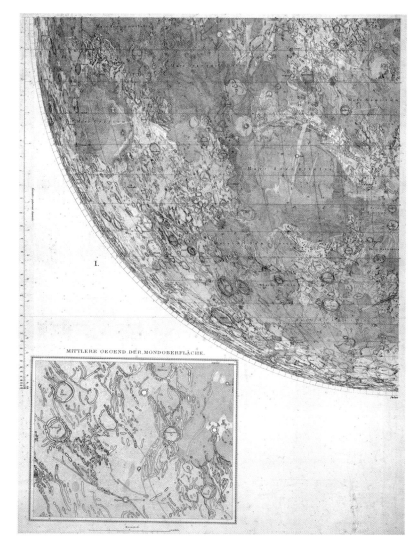

MITTLERE GEGEND DER MONDOBERFLÄCHE.

for change was slight, depending as it did on inadequately detailed earlier maps and inconsistent observations. Modern imagery shows Linné to be a perfectly ordinary impact crater about 2.45 km (1.5 miles) in diameter and surrounded by a bright blanket of ejecta material. It is a young crater by lunar standards, to be sure, but it is not as recent as Schmidt believed! Nevertheless, the controversy over

47

Linné was to rumble on well into the twentieth century, reviving interest once more in the Moon as a place were things did indeed still happen.[30]

The work of Lohrmann, Beer and Mädler, and Schmidt firmly established cartography as the dominant form of lunar science in the nineteenth century, but we should pause to consider the reasons why this should be so. It was partly driven by a belief that the Moon's violent volcanic past might not be entirely spent and that changes might continue to manifest themselves, albeit on a relatively small scale when compared to the past. Schmidt's announcement of the 'transformation' of Linné served to resuscitate lunar observation and focus it on areas suspected of change, with the result that further topographic anomalies were soon reported,

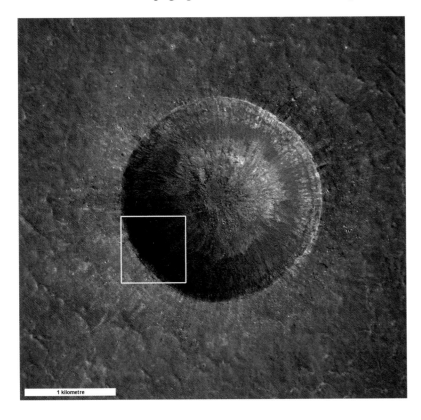

1 kilometre

Crater Linné.

Messier and Messier A observed under various angles of illumination, showing how changes in the shadows can produce the impression of changes in the shapes of the craters. South is up in these drawings by Harold Hill.

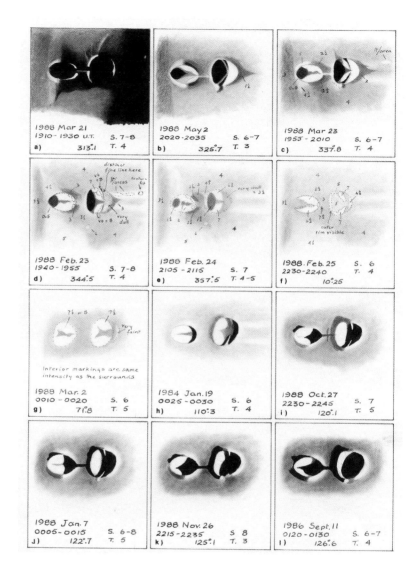

such as apparent variations in the shape of the 'twin' craters Messier and Messier A.

The dark floor of the crater Plato was also suspected of change in the form of mists and outgassings that obscured from time to time the small craterlets visible there under good seeing conditions.

Small craterlets on the floor of Plato, an area long suspected of change.

Hermann Klein's 'discovery' in 1878 of what he took to be a new crater near the Hyginus Rille also served to promote the renascent belief in lunar activity.[31]

But if such small 'changes' were to be identified with any certainty, good and highly detailed reference maps would be needed. Thus, as Sheehan and Dobbins point out, 'the stage was set for the future direction of selenographical studies: observations must focus more and more closely on minute detail. Indeed the pursuit of detail became – with a vengeance – the preoccupation of many of the selenographers who worked the field.'[32] Or, to put it another way, the *descriptive* (or cartographical) account of the minutiae of the lunar landscape came to take precedence over attempts at the geological *interpretation* of the whole, and this tendency was to shape lunar observation until well into the twentieth century. However, the obsession with detailed lunar mapping that emerged in nineteenth-century selenography may be explained in yet another way, for it was a further symptom of the essentially geographic approach to the study of other worlds that, as we have seen, had earlier been popularized by Fontenelle's *Conversations on the Plurality of Worlds*. Fontenelle's assumption that the bodies of the solar system were unknown other worlds that could be explored in the same way as our own planet found resonance in a century dominated by territorial expansionism, colonialism and imperialism. Just as nineteenth-century terrestrial explorers set out to discover and chart unknown lands and open up new worlds, so did the great lunar and planetary explorers bring the same sensibilities and assumptions to exploring alien geographies and the new landscapes they discovered in the eyepieces of their telescopes.

It is therefore unsurprising that the whole scientific approach to the study of the Moon, and to an even greater extent of Mars, came to be centred on cartography – the mapping of unknown landscapes and the exploration of new and exotic worlds. This was most clearly the case at Flagstaff Observatory, Arizona, where the businessman-turned-planetary explorer Percival Lowell devoted his life to the study of what he took to be artificial irrigation canals on Mars. Earlier, Lowell had travelled widely in the Far East and had written books on Japan and Korea. But by the late nineteenth century the one-time orientalist had turned his appetite for new worlds and remote civilizations away from the terrestrial east and out into the much more distant reaches of the solar system – and to an even more alien race that he thought was fighting for survival by building canals in the parched landscape of Mars.[33]

However, those observers who sought to bring a similar geographical approach to exploration of the Moon were, of course, dealing with much less promising material. Whereas Mars presented to Lowell and other observers a recognizably Earth-like planet replete with seasons, weather, polar ice caps and an atmosphere, the Moon appeared to offer only the craggy and apparently barren landscape of an essentially airless world. But the romantic sense of adventure and exploration persisted nevertheless, as did the corresponding expectation of new discoveries around every corner, and this fed the cartographic endeavour in the Moon's case too. So did the tendency to draw analogies between the lunar and terrestrial landscapes, a process encouraged by the familiar terrestrial nomenclature that had been attached to lunar features by earlier observers. For example, mountain ranges on the Moon had been named after such terrestrial equivalents as the Apennines, Alps, Pyrenees, Caucasus and Carpathians, and this all helped to strengthen the notion that the Moon was 'another world'.

But if by the end of the nineteenth century 'cartography had become the primary mode of representing scientific data and knowledge' about the observable other world of the solar system, as Maria Lane argues with regard to Mars,[34] then we need to take account of one essential difference between lunar cartography and how Mars and the Earth were being mapped. The widespread adoption of the Mercator projection for representing Martian topography had served to reinforce absolutely the sense of analogy with terrestrial mapping: Martian maps now began to look just like maps of our world. The map projection devised by the Flemish geographer–cartographer Gerardus Mercator, and used for his world map of 1569, was designed for nautical purposes, since it presented lines of constant course between locations. The view it presents of the world is a distorted one, particularly in high latitudes, but its purpose was to facilitate navigation by the inhabitant–explorer.

However, the Moon's captured rotation and permanently hidden hemisphere rendered the Mercator projection inappropriate for depiction of the lunar landscape. Lunar maps continued to represent the Moon's visible hemisphere from the point of view not of the inhabitant–explorer, but of the remote telescopic observer – that is, as a flat disc, complete with foreshortening at the limb and quite unlike terrestrial maps. Only with the advent of the Space Age and the imminence of first-hand exploration of the Moon did orthographic 'astronautical' maps of the lunar surface begin to appear.

On the one hand, this might have given pause to nineteenth-century observers and instilled a sense of the Moon's difference from both Earth and Mars; but on the other, it also stimulated the powerful and romantic notion of a *luna incognita*, a hidden realm on the Moon's far side. This invited flights of fancy about what might lie there, including speculation that the unseen hemisphere might be quite different from the inhospitable landscape presented by the Earth-turned side. More compellingly for the telescopic observer, it also provided an opportunity to explore those difficult limb regions revealed by libration, where mysteries possibly still abounded.

The perception of the Moon as another, as yet unexplored, territory had already long given rise to speculation that that territory, like such places on Earth, might be inhabited. The best-known example is perhaps the infamous 'Moon Hoax' of August 1835, when Richard Adams Locke used the pages of the New York *Sun* to give a fictional account of the 'great astronomical discoveries'

Examples of maps of Mars and the Moon, using different cartographic conventions.

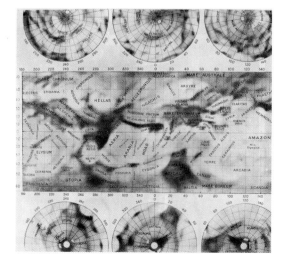

The great 'Moon Hoax'
of 1835, illustration
from New York's *Sun*
newspaper, 1835.

recently made by Sir John Herschel from the Cape of Good Hope. Those discoveries, according to Locke, included the observation of fantastic creatures disporting themselves on the banks of a lunar shore.[35]

Locke's account was a deliberate joke, but speculation about lunar life also fed the observational endeavour in a more serious way in the nineteenth century and gave rise to 'discoveries' of evidence for lunar habitation. These included observations of the remains of a lunar 'city' by the eccentric Bavarian Franz von Paula Gruithuisen in the early 1820s, as well as claims by the American William H. Pickering nearly a century later that changes in dark patches observed in and around the crater Eratosthenes might be attributable to either vegetation or the daily migration of swarms of lunar insects.

Pickering (1858–1938) was the younger (and less reputable) brother of the great astronomer Edward Charles Pickering, the director of the Harvard College Observatory who did so much to found the discipline of astrophysics. William Pickering was a fine planetary observer who for some time worked with Lowell at Flagstaff, but his observational skills were cast into the shade

by the extreme products of his imagination. Pickering advocated
a 'new selenography', one that consisted

> not in a mere mapping of cold dead rocks and isolated craters,
> but in a study of the daily alterations that take place in small,
> selected regions, where we find real, living changes – changes
> that cannot be explained by shifting shadows or varying
> librations of the lunar surface.[36]

This 'new selenography' was to find particular favour with
a British school of lunar cartography that emerged in the latter
half of the nineteenth century. Just as German selenographers
had taken the lead earlier in the nineteenth century, establishing
a tradition that was to be continued by the later work of Johann
Krieger and Philipp Fauth, so British observers now came to the
fore and were to dominate selenography until the middle of the
twentieth century. Effectively this British school began in 1862,
when John Phillips (1800–1874) presented a proposal to the British
Association for the Advancement of Science for a cooperative study
of the Moon's physical nature and the production of a 2.5-m
(100-in.) map that would improve upon that of Beer and Mädler.
The initiative had the support of the Royal Society, and in the
following year the Lunar Committee for Mapping the Surface of
the Moon was formed with William Radcliff Birt (1804–1881) as
secretary. Under Birt's guidance the committee published annual
reports between 1864 and 1869 and enlarged the scale of the
proposed map to 5 m (200 in.). But its reach exceeded its grasp and
the project was thwarted, not least by waning official support and
the cloudy skies of the British Isles. After Birt's death it was finally
abandoned, and only four of the planned 160 segments of the map
were ever completed.

In 1878 Birt co-founded the short-lived Selenographical Society
along with Edmund Neison (1849–1940), author of the first major

British guide to the Moon's surface. *The Moon and the Condition and Configurations of its Surface* (1876) was a comprehensive work of some 575 pages, containing a map in 22 sections and a full description of surface features. Both Birt and Neison took for granted the reality of physical changes on the Moon but, unlike Pickering, they ascribed those changes primarily to the likelihood of ongoing volcanic activity, rather than to the presence of life (although the possibility of some lowly form of lunar vegetation was not discounted).[37]

The case for lunar volcanism had been reinforced in 1874 by the appearance of *The Moon Considered as a Planet, a World, and a Satellite*, by James Hall Nasmyth (1808–1890) and James Carpenter (1840–1899), a work that concentrated on both the appearance of lunar surface features and what the authors called the 'causative phenomena' giving rise to them. Nasmyth was a Scottish engineer, the inventor of the steam hammer and an amateur astronomer; Carpenter was a professional astronomer working at the Royal Observatory, Greenwich. They acknowledged, as others had done before them, that there were significant differences between the craters of the Moon and terrestrial calderas, especially in scale, and they set out to explain those differences and advance a coherent theory of lunar crater formation by volcanic means. This is not the occasion to explore their results in detail; suffice to say that they saw

Nasmyth and Carpenter's volcanic fountain, from *The Moon Considered as a Planet, a World, and a Satellite* (1874).

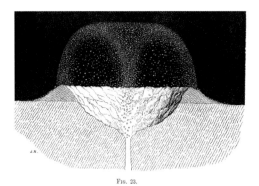

Fig. 23.

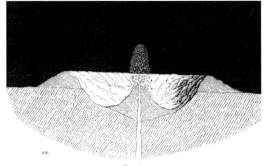

Fig. 24.

lunar craters as ring formations created as the result of a volcanic 'fountain' mechanism, whereby material that erupted from a volcanic vent fell in a ring around that vent, forming the crater walls. As the activity subsided, material vented with lesser force fell back to create a central peak at the site of the eruption.

Their theory was deeply flawed; for a start it did not fit the observed facts, especially with regard to the heights of crater floors relative to surrounding terrain. Nor was there any evidence of equivalent volcanic mechanisms on Earth that might serve as models for what was being hypothesized on the Moon. Nevertheless, their work set the pattern for subsequent British thinking on the origins of lunar craters right through to the advent of the Space Age: from Nasmyth and Carpenter in the 1870s to Moore and Cattermole in the 1960s, volcanic theories would hold sway.[38] The only significant British proponent of impact theory was Richard A. Proctor (1837–1888), better known for his book on Saturn. It was left to others to defend the impact hypothesis – most notably, the American geologist Grove Karl Gilbert and the American businessman and planetary scientist Ralph Belknap Baldwin, whose groundbreaking book *The Face of the Moon* appeared in 1949. We shall have more to say about them in the next chapter.

With the loss of the Lunar Committee and the Selenographical Society, leadership of the British school of selenography passed to the British Astronomical Association (BAA), which was founded in 1890. Its Lunar Section was directed first by Thomas Gwyn Empy Elger (1838–1897), an engineer from Bedford. He had previously served in the same capacity for the Liverpool Astronomical Society, and he was a gifted and experienced lunar observer. His notebooks reveal him to have been a cartographer and draughtsman of the highest order, and his reputation as a selenographer is undiminished to this day. He observed with a 21-cm (8.5-in.) reflector and produced in 1895 a book containing a 45-cm (18-in.) map that is still useful to the observer.[39]

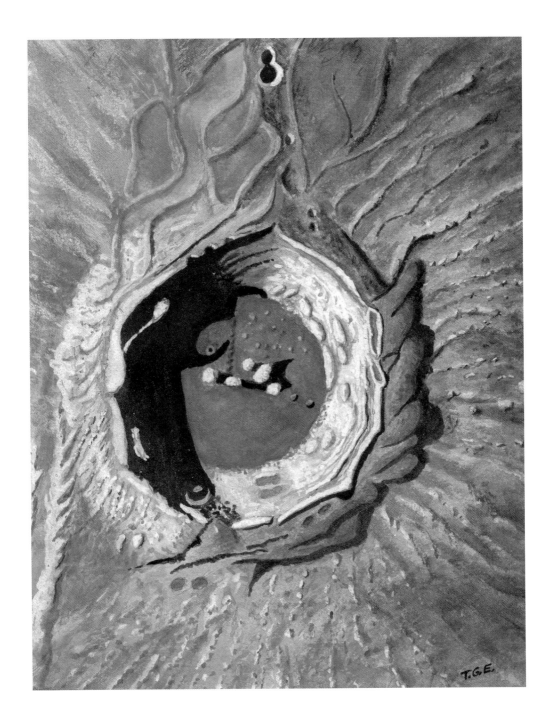

Although essentially a lunar cartographer in the traditional mould, Elger did go beyond the mere representation of surface detail in order to speculate about the nature and origins of lunar features. Considering the age in which he worked, some of his ideas were reasonable. For example, he argued that mare wrinkle ridges were not the results of alluvial action on the beds of former seas, as some still contended, but were instead the products of volcanic action and lava flows.[40] He also realized that the bright ray systems associated with certain craters were ejecta deposits. However, he failed to recognize impact as the mechanism giving rise to those deposits, preferring instead to see them as the result of volcanic eruptions.[41]

In the end, though, Elger was a man of his time, and his selenography was driven by the same imperative of pursuing ever more detailed maps that would settle once and for all the vexed question of changes and activity on the Moon. As he wrote in his book:

> the more direct telescopic observations accumulate, and the more the study of minute detail is extended, the stronger becomes the conviction that in spite of the absence of an appreciable atmosphere, there may be something resembling low-lying exhalations from some parts of the surface which from time to time are sufficiently dense to obscure, or even obliterate, the region beneath them.[42]

Elger's successor, Walter Goodacre (1856–1938), was the longest-serving director of the BAA Lunar Section. In 1931 he privately published his book *The Moon*, containing an excellent 195-cm (77-in.) map,[43] but his approach offered little that was new and he continued to emphasize the discovery of ever-finer detail in a way that clearly reaffirmed cartography as the heart of observational activity. In 1933 he wrote:

Crater Copernicus, 1889, drawn by Thomas Gwyn Empy Elger.

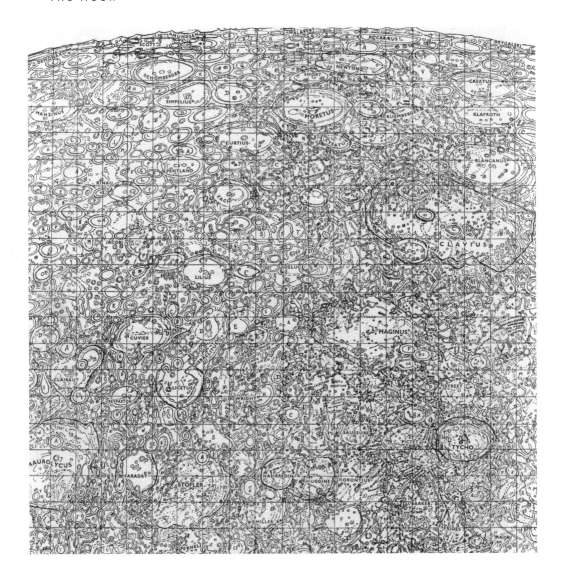

One of the chief sources of pleasure to the lunar observer is to discover and record at some time or other details not on any of the maps; it also follows that if in the future a map is produced which shows all the detail visible in our telescopes, then the task of selenography will be completed.[44]

A section of H. P. Wilkins's 7.5-m (300-in.) map.

60

By the mid-twentieth century, however, the writing was on the wall for the great tradition of lunar cartography started by Harriot. The development of bigger and better telescopes had indeed disclosed finer and finer details on the Moon's surface, but it had also revealed how the charting of that detail was turning into an endless and pointless task. Maps made by telescopic observers simply became bigger and bigger and ever more cluttered by confusing detail. The Welshman Hugh Percy Wilkins (1896–1960) was perhaps the last of the great amateur Moon-mappers, producing in succession charts of 2.5, 5 and 7.5 m (100, 200 and 300 in.) in diameter. All were driven by the pursuit of ever-smaller detail, and it is with them that we see the unfortunate but inevitable results of that approach. The maps – especially the one that measures 7.5 m – are overcrowded to the point of being indecipherable.

Moreover, the Space Age and first-hand investigation of the Moon by spacecraft were just around the corner – indeed, Wilkins's maps were used by the Russians to interpret the images of the lunar far side returned by Luna 3 in 1959. But that acknowledgement of past achievements in selenography was to prove valedictory: the future of lunar science was about to leave behind both the amateur Moon-mapper and the great age of telescopic observation.

THREE

The Moon in the Modern Age

As the nineteenth century gave way to the twentieth, the shape and direction of lunar science began to shift, at first slowly and then with bewildering speed, eventually producing results that would have been inconceivable to scientists in 1900. That process emerged primarily from three new factors. The first was the development of photography to the point where it became a viable and effective tool for astronomical research. The second was the professionalization of astronomical science. And the third was a period of rapid technological development that within a single century saw humankind progress from the invention of the aeroplane through to manned Moon landings and the robotic exploration of all the major worlds of our solar system.

Photography, of course, had been around since the experiments of Nicéphore Niépce and Louis Daguerre in the 1820s and '30s. However, early techniques lacked the sensitivity necessary for the capture of faint or moving celestial objects, so the development of photography and astrophotography did not move forward at the same pace. John W. Draper is credited with producing the first clear photograph of the Moon in 1839, but the amount of detail it showed was limited. Between 1858 and 1862 Warren de la Rue used his 33-cm (13-in.) reflector at Cranford, Middlesex, to take stereoscopic images of the Moon. His telescope was clock-driven to compensate

Full Moon by
Warren de la Rue.

for the rotation of the Earth, and this permitted the long exposures necessary to capture worthwhile images.

Following this, technological progress was sufficiently rapid for W. H. Pickering to be able to use images taken in the first year of the twentieth century as the basis for a photographic atlas of the Moon, published in 1904 as part of his well-known book *The Moon*.[1] The photographs were taken with a 30-cm (12-in.) telescope stopped down to an aperture of 15 cm (6 in.) and they provided reasonably sharp images of parts of the lunar surface under different angles of illumination. However, they also showed the limitations of photography when it came to trying to reconcile the long exposures it demanded with the unsteady state of the Earth's turbulent atmosphere. The result was that fine detail was smeared out and Pickering's photographs revealed little more than could be seen visually in a 5- or 8-cm (2- or 3-in.) telescope.

A more systematic attempt at a photographic atlas was made between 1896 and 1909 at Paris Observatory, using the 61-cm (24-in.) Coudé refractor, but the results offered little improvement in resolution.[2] Photographs did, however, provide a basis for more accurate positional and measurement work on the Moon's major features, and both Julius Franz (1847–1913) at Breslau and Samuel A. Saunder (1852–1912), a British amateur, used photographs to produce separate catalogues of lunar coordinates in the first decade of the twentieth century. But even when the world's largest telescopes, such as the Mount Wilson 254-cm (100-in.) reflector and the 91-cm (36-in.) refractor at Lick Observatory, both in California, started taking lunar images in the 1920s and '30s, the observational advantage remained with the visual observer, who could detect much finer detail in fleeting moments of sharp seeing. One observer who successfully brought together photographic and visual techniques was the Bavarian amateur Johann Krieger (1865–1902). His beautiful *Mond-Atlas*, incomplete but published posthumously in two volumes in 1912, used low-contrast photographic prints, mainly from the Paris atlas, as a base upon which Krieger sketched finer detail seen through the eyepiece of his 27-cm (10.5-in.) reflector.[3]

The first photographic atlas to combine a systematic approach with reasonably high resolution was that produced in the 1950s by the great planetary scientist Gerard Kuiper (1905–1973).[4] As director of the Yerkes Observatory in Wisconsin and the McDonald observatory in Texas, Kuiper was able to avail himself of the best professional photographs taken with some of the world's finest telescopes, and his *Lunar Atlas*, published in 1960 as a boxed set of 230 sheets, was to remain unmatched until the advent of digital imaging. It became the benchmark of pre-Space Age lunar photography and it is still a useful reference today.

Drawing of Gassendi, from Johann Krieger, *Mond-Atlas* (1912).

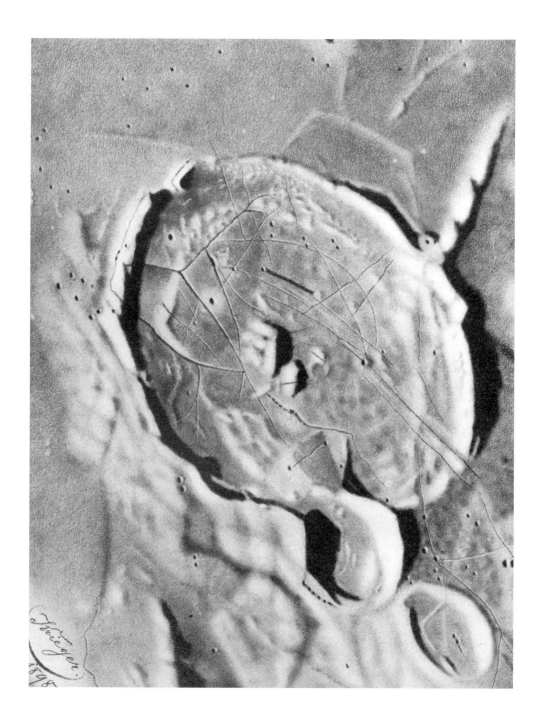

But Kuiper's achievement was significant in yet another respect, for it marked the moment when leadership in lunar science passed finally from the amateur to the professional. Prior to the twentieth century lunar observation had been almost exclusively the preserve of the dedicated private enthusiast – often a 'grand amateur' of independent means, equipped usually with only modest equipment but rich in time and enthusiasm. Indeed the concept of the professional astronomer was a tenuous and ill-defined one before the twentieth century. Even then, most professionals were drawn to the new developments in astrophysics, as the subject itself focused upon understanding the large-scale universe rather than the backyard of our own solar system. The Moon was an object of particular neglect, regarded as parochial and of little importance when compared with the great strides in extra-galactic astronomy being made by figures such as Edwin Hubble (1889–1953). Hubble was the first man to recognize that many of the so-called nebulae were in fact other galaxies outside the Milky Way – island universes in their own right.

Despite that, the Moon had not been entirely ignored by professionals before Kuiper. The International Astronomical Union, founded in 1919, had set up a commission to regularize lunar nomenclature. The body drew upon work done earlier by Mary Blagg (1858–1944), an English amateur who had collaborated with Saunder and was one of the first women to be admitted as a Fellow of the Royal Astronomical Society. In 1935 the commission published its results, consisting of a catalogue and maps of lunar formations.[5]

However, it was Kuiper who saw that, with the advent of the Space Age and the possibility of manned exploration of the Moon in the not-too-distant future, sustained professional investment in lunar cartography and lunar science would pay handsome dividends. In 1960 he moved to the University of Arizona to found the Lunar and Planetary Laboratory (LPL), working with a group of younger colleagues and graduate students. These included Ewen

Whitaker, who went on to become a world-renowned historian of selenography and a major player in the planning of later NASA missions to the Moon. Whitaker had previously been an amateur selenographer and had served as director of the Lunar Section of the British Astronomical Association. He joined Kuiper's fledgling lunar group at Yerkes in 1957, where he participated in the lunar mapping project. Whitaker's recollection of how he became part of Kuiper's team is a measure of the lack of professional interest in the Moon at that time:

> It really dates back to 1955, when I went to a meeting of the International Astronomical Union in Dublin, Ireland. I'd heard that Gerard Kuiper was going to be there, which was fortunate because I knew that he was interested in the Moon and planets. He put out this little memo: 'I'm interested in making an atlas with the best pictures that have been taken from the Mt. Wilson, the Pic-du-Midi observatories, and anyone interested in giving some guidelines, please get in touch after the meeting.'
> Well, I got back home and I thought, 'This is good, I'm interested in the Moon as a sideline, let me write to Kuiper.' So I wrote him a long letter in longhand . . . and of course he wrote back. I was the only one who wrote to him out of all the astronomers at that meeting. Four hundred astronomers, but not one was interested in the Moon. I was the only one.[6]

Kuiper and his team embarked upon an ambitious programme of lunar research that would lay the groundwork for later missions to the Moon. Apart from further systematization of lunar craters and the production of orthographic and rectified lunar atlases, the work done at LPL paved the way, directly or indirectly, for the Lunar Aeronautical Charts drawn up by the U.S. Air Force and the geological and stratigraphic maps produced by Eugene Shoemaker and others at the U.S. Geological Survey. The former charts reversed the previous 'astronomical' convention of east and west on the Moon, which corresponded to celestial east and west as seen in the sky from Earth, in favour of one designed to meet the directional expectations of future astronauts trying to find their way around the surface. The latter maps brought together, for the first time properly in the case of the Moon, the disciplines of cartography and geology, which had so often taken conflicting paths in the past.

Impact Theory takes Centre Stage

The work of the LPL was also revolutionary because of the compelling evidence it accumulated in favour of the impact theory of lunar crater formation. As we have seen, most lunar observers throughout the centuries since the invention of the telescope had come out in favour of lunar volcanism as the primary mode of lunar surface formation. The possibility that the craters of the Moon might have been produced by asteroid and comet impacts was largely dismissed. To a certain extent this was in keeping with the amateur status and enthusiasms of most of those telescopic observers. They needed something to draw them back to the eyepiece night after night, but impact theory inevitably implied an essentially dead Moon, a celestial museum displaying the scars of a violent but long-gone past. Volcanism, on the other hand, held out at least the possibility of continuing change and activity, the potential for seeing something new at each observational opportunity – a much more alluring prospect.

Ironically, Kuiper's own interest in the Moon had first been fed by volcanism and his recognition that the lunar maria had to be the results of extensive lava flows, albeit from an era long past when magma from the Moon's interior, molten by radioactive decay, spilled out onto the surface. However, one of Kuiper's graduate students came to realize that the surface structures into which those lavas had flowed and solidified must have pre-existed and been formed by an entirely different process. William K. Hartmann (b. 1939) came to LPL in 1961 and worked on producing rectified views of the lunar limb regions. He describes how the building in which he worked

> had a tunnel in which we projected photographs of the Moon onto a globe – onto actually a half-globe, a three-foot, white half-globe – and then re-photographed that globe from different directions so that we could see the structures on the Moon as they would look from overhead.
>
> When we projected images on that globe, we could walk around to the side and see these structures in ways that people had really never seen before. We discovered that, particularly, there was a big, beautiful bulls-eye structure, a multi-ring basin that turned out to be an impact structure – huge, a thousand kilometers across – on the east [now west] limb of the Moon.

The Orientale multi-ring impact basin.

It's called the Orientale Basin. Looking at that made it obvious that a lot of the other basins, like the Imbrium basin and Nectaris and so forth, were the same class of multi-ring bulls-eye structures. We could trace these rings.[7]

Telescopic observers had long been aware of a patch of marial lava that had been given the name Mare Orientale to denote the fact that it lay right on the Moon's then eastern (now western) limb. They had also observed that the Mare Orientale was edged by mountains, which were named the Rook and Cordillera ranges. What had not been recognized until Hartmann's experiment was that those mountains formed the rings of a huge basin in which the mare lavas lay, and that such a huge single structure could not have been formed by any forces associated with volcanic action. It had to be an impact structure.

Some twenty years later Hartmann reflected on the significance of his discovery in a seminal paper. He described how he came to understand that the key to unlocking the secrets of the Moon's surface had been hidden in plain sight all along. The devil had indeed been in that very detail that observational tradition had insisted upon plotting so obsessively, rather than trying to recognize patterns in the large-scale structures seen on the Moon. Hartmann's paper advocated what he termed 'gestalt perception', whereby an atomistic outlook is replaced by recognition of 'an organised whole that is perceived as more than the sum of its parts':[8]

The individual elements of the pattern were visible to generations of observers and indeed were mapped and discussed in minute detail. The observers, however, did not perceive from these studies that the moon's largest craters are multiple ring systems of concentric and radial elements in widely repeated patterns. These facts have some importance, perhaps, in our conduct of scientific research, since they show that intimate knowledge of relevant details does not insure recognition of 'big pictures'.[9]

Hartmann described the moment when the rectified images revealed the Orientale basin to be 'an enormous system of elegant symmetry' as a 'eureka experience' for himself.[10] But he was also quick to recognize that others before him had come to understand both the importance of a holistic view and the role played by impact in sculpting the lunar surface. Grove Karl Gilbert (1843–1918), one of the greatest American geologists and a co-founder of the U.S. Geological Survey, had given an address to the Philosophical Society, Washington, on 10 December 1892. Whitehouse describes that address as 'the beginning of [a new] stage in the Moon's shifting allegiances', away from the astronomers and cartographers and towards the geologists.[11] A field study of Coon Butte crater, Arizona, (now better known as 'Meteor Crater') in 1891 had prompted Gilbert's interest in the craters of the Moon, although ironically he eventually concluded wrongly that the Arizona crater was of volcanic origin. Geology was a relatively new science in the late nineteenth century and Gilbert's experiences at Coon Butte had brought home to him the fact that the Moon was another 'rocky world' awaiting exploration using the same investigative methods.

This was the start of a process whose culmination is marked in the title of Don Wilhelms's book *To a Rocky Moon: A Geologist's History of Lunar Exploration* (1993). Wilhelms's mammoth work appeared when the age of lunar exploration by spacecraft and manned missions was already well under way, and his title eloquently confirms the final annexation of our Moon by geologists and geological science:

The Moon did once belong to astronomy, the study of distant reaches where humans have not yet gone . . . The approach of lunar exploration during the 1960s, however, destined the Moon to become not only a globe to be measured and tracked, or a

surface to be scanned by instruments, but also to become known as a world of rock. Lunar science increasingly became geological science. The later Apollo missions were elaborate geologic field trips.[12]

Gilbert pursued his less elaborate lunar 'fieldwork' using the 66-cm (26-in.) refractor at the U.S. Naval Observatory, Washington, and he quickly reached the conclusion that lunar craters were completely different in structure from terrestrial volcanoes and could not have had a volcanic origin:

> [The] greater maximum width [of lunar craters] constitutes a real difficulty, especially as volcanoes appear to have a definite size limit, while lunar craters do not. Form differences effectually bar from consideration all volcanic action involving the extensive eruption of lavas . . . The volcanic theory, as a whole, is therefore rejected.[13]

But if volcanism was out as an explanation for the creation of craters displaying the size and morphology of those on the Moon, many aspects of lunar crater structure *did* correspond to characteristics of the scars left when projectiles were fired into pasty or plastic materials. As a result, Gilbert came down firmly in favour of the impact hypothesis, even though he acknowledged that there were objections still to be overcome. For a start, what sort of projectile could have given rise to such enormous structures? Also, if the Moon bears so many impact scars, why does the Earth appear to have so very few? Why were the overwhelming majority of lunar craters round, when most incoming projectiles would presumably have impacted at oblique angles and given rise, surely, to eccentrically shaped craters?

Ejecta 'sculpture' around Fra Mauro, as seen from Apollo 16.

These were questions that would be answered fully only by later scientists, but Gilbert's analysis did get many things right. He was able to account for the bright ejecta rays surrounding craters such as Tycho and for the terracing found inside many larger formations. He also showed how the impact mechanism could produce central peaks in craters, features long held previously to be volcanic cones. But, most tellingly, Gilbert's gestalt perception allowed him to step back and recognize large-scale patterns of radial grooves and furrows, apparently emanating from common points of origin, patterns that he described as surface 'sculpture'.

In particular, he used those patterns to identify the likelihood of 'a collision of exceptional importance' that had occurred in the Mare Imbrium and had resulted in 'the violent dispersion in all directions of a deluge of material – solid, pasty, and liquid'.[14] Even a small telescope will show widespread evidence of that dispersed matter, particularly to the south of Imbrium and towards the centre of the lunar disc. For example, the conjoined craters Fra Mauro, Parry and Bonpland are scoured by such material, and the western floor of Fra Mauro itself is covered in hummocky ejecta deposits thrown out by the Imbrium impact. Apollo 14 was sent to that area especially to sample the deposits, determine their age, and confirm that they were indeed composed of the shocked, melted and brecciated material one would expect to find in ejecta from a catastrophic impact.

The next truly significant advance in impact theory was made by another American, Ralph Belknap Baldwin (1912–2010), a man of great intellectual agility and wide experience who 'looked at the moon with a new eye' and 'introduced lunar science to the twentieth century'.[15] His book The Face of the Moon (1949) has stood as an inspiration to subsequent generations of lunar scientists, and in an age before speculation about the Moon could be tested against results returned by spacecraft, it essentially got most things right. An 'astrophysicist by education, industrialist by profession, and versatile lunar scientist by avocation',[16] Baldwin earned his living in his family machinery business, but he also taught astronomy whenever he could. It was while waiting to give a lecture at the Adler Planetarium in Chicago that his interest in the Moon was piqued by a display of lunar photographs showing the patterning around the Mare Imbrium that Gilbert had described as lunar 'sculpture'.

Baldwin was unaware of Gilbert's work and he was to remain so until 1948, well after his own work for *The Face of the Moon* was essentially completed; but like his predecessor, he saw such sculpture as clear evidence of a catastrophic impact that had created the Imbrium basin. However, Baldwin had studied bomb craters both during and after the Second World War, and he quickly recognized that the craters and basins of the Moon had to be the results not only of impact, but of an immense explosive force. They could not have been produced by volcanic means such as updoming or intrusions of magma, or indeed by 'any other nonexplosive method'.[17] Instead, argued Baldwin, 'the case for the explosive origin of the moon's craters is unassailable. The probability is very great that the explosions were caused by the impact *and sudden halting* of large meteorites.'[18] The immense size of lunar basins and craters could be explained only if we were to fully understand 'the physical processes which occur when a rapidly moving body strikes a solid surface'.[19]

Those processes do not involve the simple excavation of a crater by an impactor. Instead the crater is the result of the immense explosive energy released when an asteroid, comet or meteoroid, perhaps up to many kilometres in diameter and travelling at interplanetary velocities greatly exceeding anything we can experience (or even imagine) on Earth, is suddenly arrested by impact with the solid surface of the Moon. We are dealing with forces far in excess of those imagined by most earlier advocates of impact theory, who thought in terms of dropping pellets, or even firing bullets, into pliable clay. Baldwin acknowledged the earlier work of the New Zealander Algernon Charles Gifford (1861–1948) in recognizing the immense violence involved in the collision of interplanetary bodies speeding in rapid orbits around our solar system. The Earth, in its own sedate and ordered revolution around the Sun, is travelling at some 30 km per second (67,000 mph). The Chelyabinsk bolide of 2013, which exploded over Russia, causing considerable damage to buildings, entered Earth's atmosphere at a speed of around 19 km per second. But an asteroid or comet sweeping in towards perihelion, or closest approach to the Sun, from the outer reaches of the solar system could reach speeds far in excess of that. Moreover, the lack of an appreciable atmosphere on the Moon means that the speed of an incoming massive body is not slowed, as it was in the case of Chelyabinsk, and so the kinetic energy at impact is undiminished.

This was the starting point for Gifford's argument that 'a meteorite moving at a speed of many miles a second has a store of kinetic energy which renders it, when suddenly stopped, an explosive of unparalleled violence.' Indeed the energy of an impact is determined by the mass of a projectile and the square of its velocity. On impact a small percentage of the energy released goes into vaporizing the impactor, but, according to Gifford, that is 'a negligible fraction of that which it possesses' and the rest goes into compacting, shattering and pulverizing the surrounding rock, hollowing out a 'great saucer-shaped depression' and building the ejected material into a raised rim.[20] Owing to the explosive forces released by the impact, the size of the impactor may be much smaller than the crater it gives rise to. Interestingly, a similar idea had been floated at a meeting of the British Astronomical Association on 26 May 2015 by Professor A. W. Bickerton. The meeting report reads:

[Professor Bickerton] thought that in all probability a meteor, striking the surface of the Moon and going in for some distance, would produce volcanic action . . . Thus, although the primary cause of the craters in the Moon might be due to impact, the real force that produced the crater was due to volcanic action. That, he thought, was a possible interpretation of their circular form.[21]

Bickerton's choice of the term 'volcanic action' is unfortunate, for he is clearly referring to *explosive* action that releases molten rock, but his recognition that such action might explain the circular form of lunar craters is of fundamental interest. As we have seen, one of the most telling objections to earlier impact theory had been the argument that impactors arriving at an oblique angle would produce eccentrically shaped craters, but the processes envisaged by Bickerton, Gifford and Baldwin relied upon what was effectively a sudden point-source explosion of great magnitude to produce circular craters around the site of the explosion almost regardless of their angle of arrival.

Crater Archimedes, showing how later lavas have encroached upon the outer walls and filled the interior. The small pits on the smooth floor are later still.

If Baldwin found the source of his thinking in the ideas of Gifford, then he subsequently succeeded in developing those ideas into a far grander and more coherent explanation of the face of the Moon. Perhaps most importantly, he collated a mass of data that made the case for an explosive origin of lunar craters unassailable, by showing the regularity of depth–diameter relationships among terrestrial explosion craters, terrestrial meteorite craters and fresh, unmodified lunar craters (that is, those in a relatively pristine state, rather than those degraded by later impacts or ejecta deposits). All these craters were thus of the one family and were produced by the same mechanism:

> There is thus a very smooth curve which represents equally well the largest [unmodified] lunar crater and the smallest

terrestrial explosion pit. These two groups, tied together perfectly by craters [on Earth] of known meteoritic origin, form a relationship which is too startling, too positive, to be fortuitous . . . The only reasonable interpretation of this curve is that the craters of the moon, vast and small, form a continuous sequence of explosion pits, each having been dug by a single blast. No available source of sufficient energy is known other than that carried by meteorites.[22]

Moreover, Baldwin recognized that the circular maria were simply larger versions of craters, produced by the same explosive mechanism but involving impactors of much greater size. Gifford had also considered this but Baldwin, crucially, drew a distinction between the mare basins and the lava flows that filled them. In Gifford's view both were products of the same event: the force of the impact that created the basins also unleashed liquid molten material from beneath the lunar crust that 'deluged a vast circular area with fiery spray'.[23]

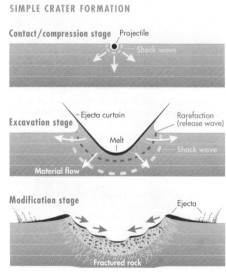

A simple crater and its process of formation.

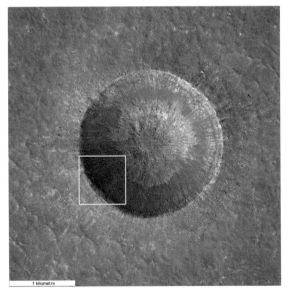

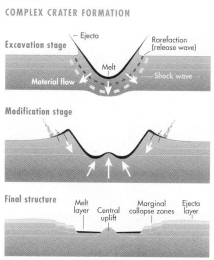

COMPLEX CRATER FORMATION

Excavation stage — Ejecta, Rarefaction (release wave), Melt, Material flow, Shock wave

Modification stage

Final structure — Melt layer, Central uplift, Marginal collapse zones, Ejecta layer

A complex crater and its process of formation.

However, Baldwin saw that the dark lavas that comprised the maria themselves were the results of a separate and much later process, as the lavas that now filled the basin floors had breached, buried or overflowed craters that must have formed after the impact basin itself. A prime example of this is the crater Archimedes in the Mare Imbrium. The fact that this crater formed on the floor of the Imbrium basin means that it was created later, and the subsequent enclosure of Archimedes by mare lavas and the flooding of its floor means that the infilling lavas must be younger still. This 'superimposition' of one feature on top of another was thus recognized by Baldwin as an effective way of establishing the relative ages of lunar surface features and it is still a valuable technique today.

Thanks to the insights and elegant arguments provided by the likes of Gilbert, Gifford and Baldwin, we now have a fairly complete understanding of the morphology of lunar impact craters and the processes involved in their formation. Although those processes are generally similar for all impact craters great and small, as Baldwin showed, there are important modifications as we progress in size from the smallest pits to the largest. Smaller impacts, producing craters up to about 15 km (9 miles) in diameter, involve a relatively simple (albeit still violent) process, consisting of an excavation stage followed by a modification stage. The high-speed impactor compresses the lunar rocks at the point of arrival, pushing material downwards and squeezing it outwards. The shock waves and immense energy released instantaneously on impact shatter and melt the rock and vaporize the meteorite. The shock waves are driven downwards then rebound, ejecting further material from the infant crater and forming the outer rim from material excavated in the impact. Finally, as the shock waves subside, fractured rock settles to form a bowl-shaped pit with a raised rim, a floor below the level of the surrounding terrain, and debris slides or scree on the inner walls. As Charles Wood nicely puts it, for the simple crater, 'life's one great fling is pretty much over' at that point.[24]

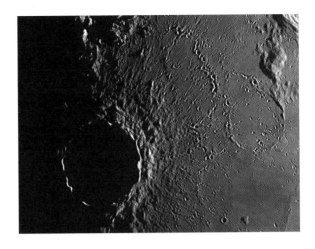

Ejecta blanket and secondary cratering near Copernicus.

The creation of larger craters, say in excess of 40 km (25 miles) diameter, involves larger and/or faster impactors and the release of much more destructive energy, and this has the effect of complicating the morphology of the craters produced. Again, it is a two-stage process, consisting of excavation of the crater followed by modification. It begins, as in the creation of simple craters, with the compression of rock at the point of impact and downward shock waves that eventually rebound. But now the rebound forces are so great that they dredge up material from deep below the newly formed crater to form a central peak or peak complex. Moreover, such is the size and depth of the excavation that the inner walls are unstable and unable to resist the force of gravity. Instead they

The intermediate crater Dawes.

collapse into the spectacular concentric slumps of terracing that we see inside craters such as Copernicus, Theophilus and Tycho. Finally, rocks melted by impact pool to cover the crater floor, so that the overall profile is shallower and flatter than the bowl-shaped profiles of simple craters.

Moreover, the additional violence involved in the creation of a large complex crater also generates an appreciable blanket of ejecta over the surrounding terrain. This takes the form of an ejecta layer,

or glacis, beyond the rim and the scarring of the adjacent area by
secondary cratering caused by blocks of material thrown out from
the primary impact site. This is particularly evident on the mare
surface around Copernicus.

In between the simple and complex craters we find others with
an intermediate type of morphology. These usually range between
15 and 35 km (9 and 22 miles) in diameter and they may or may not
have central peaks or small central mounds. However, they do show
clear evidence of structural failure in their walls, which have slumped
in places to produce floor debris and scalloped rims, but without
forming the complete concentric terracing of their larger brethren.
A fine example is crater Dawes, 18 km (11 miles) in diameter, located
at the junction of the Mare Serenitatis and the Mare Tranquillitatis,
where partial wall collapses have resulted in a scalloped rim and
irregular slumps of material on the floor.

Just as there is no precise size at which a simple crater becomes
an intermediate one, or the latter becomes a complex crater, so
there is no clear cut-off point in size at which a large crater becomes
an impact basin. But both landforms belong to the same 'main
sequence' of impact cratering. Baldwin acknowledged this when
he used the term 'supercraters' to describe the lunar maria, going
on to say: 'Such a correlation means that the mode of origin of these
seas is the same as that of the craters and hence that all parts of their
observed structures were caused by the impact of vast meteorites,
one per mare, and resultant reacting processes on the moon.'[25]

However, as Paul Spudis has shown, with increasing size we
do see a developing morphology that reflects differences in the
impact mechanics brought on by the increased energies involved.
Large complex craters with central peaks seem to occupy a size
range approximately between 40 and 200 km (25 and 124 miles)
in diameter. It is not possible to be more accurate than this since
the great age of most craters at the larger end of this range means
that they have been subjected to significant modification by ejecta

The Schrödinger Basin,
showing an inner peak
ring instead of a central
peak. This is a Lunar
Reconnaissance Orbiter
(LRO) image with a false
colour overlay showing
gravity variations.

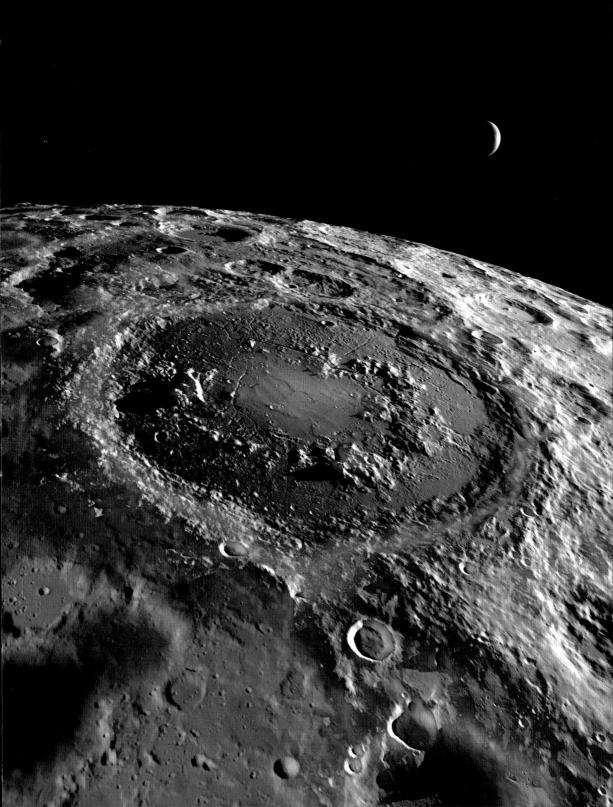

damage from later impacts in their area, so that we cannot see their pristine states. Nevertheless, Spudis identifies several intermediate forms between large complex craters and fully formed multi-ringed impact basins such as Orientale.

As we have seen, complex craters display central peaks created by the rebound of rock from the compression of impact. As we approach the top end of the size range for complex craters we encounter formations that display not only a central peak complex and wall terracing, but indications of rings of individual peaks between the crater centre and walls. Spudis offers the far-side crater Compton (162 km (100 miles) in diameter) as an example of such a protobasin,[26] and imagery from spacecraft clearly shows an incipient multi-ring structure surrounding the central peaks.

Further up the size-scale, somewhere between 250 and 300 km (155 and 186 miles) in diameter, we encounter the transition to true basin structures that show much clearer evidence of a secondary inner ring, but this time *instead of* a central peak. Schrödinger, 312 km (194 miles) in diameter, is a fine example, albeit again on the Moon's averted hemisphere and close to the south pole. Finally, we see fully formed impact basins such as Orientale, which show evidence of more than two concentric ring structures, both within and outside the primary rim.

It is now recognized that impact basins are the largest topographic structures not just on the Moon, but throughout the solar system, and they have been found on Mercury, Mars and other solid bodies. Yet the mechanism by which central peak craters evolve with size into peak-ring structures, and then into true multi-ring basins, remains unclear. Various theories have been advanced, including a 'tsunami wave' model that attributes the ring structures to the movement of rocky material fluidized

The Compton 'protobasin', showing an incipient ring-structure on its floor.

by impact, similar to the ripple-rings produced when a stone is cast into water. Alternatively, basin inner rings have been described as complex rebound structures, where the uplift of material is followed by subsequent collapse, and outer rings have been ascribed to crustal deformation and undulation in response to the initial impact, creating concentric fractures and collapse scarps. The jury is still out![27]

Volcanic Theory Fights Back

It is a tribute to Ralph Baldwin's achievement that he identified much of the morphology of impact craters and multi-ring impact basins. This included recognition of isolated peaks on the surface of the Mare Imbrium as possible remnants of a submerged inner ring. He also anticipated many features of the mechanisms involved, such as arguing that craters' central peaks were rebound structures and not volcanic cones as many had held. Yet, despite the force and elegance of his arguments, adherents of volcanic theory were still prepared to fight a stubborn rearguard action even into the age of direct exploration of our Moon by spacecraft. *The Face of the Moon* had answered so many of the objections raised earlier against impact theory by the volcanists, including, most importantly, the question of why impacts by bodies coming in from different and oblique angles did not produce many more non-circular craters. But other objections still remained.

For much of the twentieth century, before the Space Age, disagreements between the two camps reflected two great divides. The first was between amateur observers, who were usually unversed in geological theory but favoured volcanism for the reasons we have discussed earlier, and those few professionals who took an interest in the Moon, who were more alert to the geological differences between lunar craters and terrestrial volcanic calderas. The second was strangely geographical: European observers largely seemed to prefer volcanism, while impact theory gained more support in the United States.

One American geologist who did break this pattern, and argued emphatically for a volcanic mechanism for crater formation, was Josiah Edward Spurr (1871–1950), although it is fair to say that his conclusions were embraced largely by Europeans and ignored by his fellow Americans. Spurr had identified the same matrix of lineaments crossing the lunar surface that Gilbert had called 'sculpture'. However, whereas the word 'sculpture' implied something that had been carved out and was used by Gilbert to describe secondary impact scars and ejecta deposits, Spurr preferred the term 'grid system'. He saw this as evidence that the Moon's surface features were not randomly distributed, as you would expect from the impact hypothesis, but were instead arranged along lines of tectonic faulting in the lunar crust. This faulting marked sites of crustal tension and weakness, and provided a locus for eruptive volcanism.

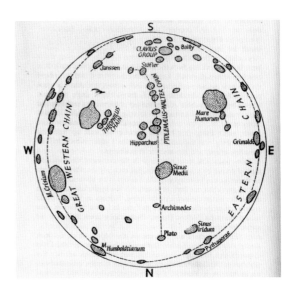

Non-random distribution of lunar craters, according to Patrick Moore, from *Survey of the Moon* (1963).

In the mid-1960s Spurr's tectonic grid system found favour with the British authors of two books that turned out to represent the equivalent of Custer's last stand as far as volcanic theories of crater formation were concerned. These were Gilbert Fielder's *Lunar Geology* (1965) and *The Craters of the Moon* (1967) by Patrick Moore and Peter Cattermole. Fielder and Cattermole were professional geologists, while Moore (1923–2012) was an amateur lunar observer of long experience whose work as perhaps the most widely known popularizer of astronomy in the twentieth century meant that his opinions carried considerable force. Fielder came to embrace volcanic theory only after a long flirtation with impact

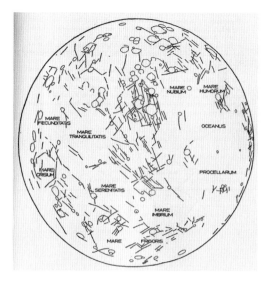

Moore and Cattermole's representation of the lunar 'grid system'.

theory, but Moore was hostile to the latter from the start and his eventual conversion after the revelations of the Apollo programme was effected with great reluctance. In 1949, upon the appearance of Baldwin's classic, he wrote: 'Have just got hold of the new book by Baldwin, *The Face in* [sic] *the Moon*. I haven't read it but am prepared to disagree with every word – I have no time for the meteor theory.'[28]

Despite such bullishness, Moore's views were deeply and honestly held. They were based on interpretation of his own observational experience and always compellingly argued. He often expressed the view that anyone who took the trouble actually to look at the lunar surface through a telescope would quickly become convinced that volcanism was the main agent of crater formation. This was distinctly unfair on someone like Baldwin who, although not a telescopic observer, nevertheless derived his arguments for impact theory from a sound knowledge of lunar topography.

Moore's objections to impact theory were essentially those that had been expressed by volcanists in the past. In essence, he was struck by what he saw as the non-random distribution of lunar surface features. In particular, there were many examples of craters apparently arranged in grand arcs and chains which he charted and named.

For adherents of volcanic crater theory such alignments were incompatible with impact theory, since any bombardment should surely result in a random distribution of craters. Much more likely, they argued, craters had formed in a patterned way as a result of volcanic forces acting along lines of crustal weakness. For Baldwin and his supporters, however, such alignments were indeed either

Overlapping craters, Thebit.

chance in nature or merely a trick played on the eye by the effects of foreshortening in the limb regions. They were not part of any 'grid system', as first identified by Spurr and subsequently described in detail by later writers such as Fielder, who detected what appeared to be an elaborate network of slip/strike faults, horsts, rilles, wrinkle ridges, crater chains and other linear features that structured the entire face of the Moon.

The Davy crater chain and the fragments of Comet Shoemaker–Levy 9 that impacted Jupiter in 1994.

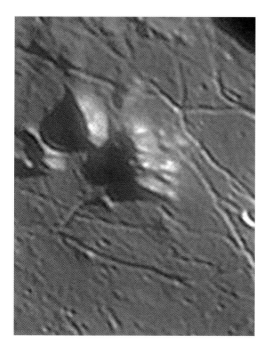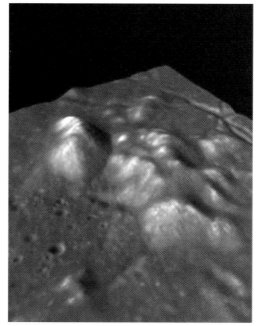

Left: apparent summit pits in Gassendi; right: QuickMap LRO 3D model, showing the 'molar' shape giving rise to the impression of summit pits.

Moore saw similar order in the way in which craters overlapped, for it is true that in the majority of cases where one crater intrudes upon another it is the smaller crater that interrupts the larger. The example most often quoted is that of the crater Thebit, 57 km (35 miles) in diameter, on the eastern edge of the Mare Nubium. Thebit is interrupted by Thebit A (20 km (12 miles) in diameter), which is in turn broken into by Thebit L. (10 km (6 miles) in diameter) Moore's argument was that such a pattern, replicated over the entire lunar surface, was wholly explicable in terms of a gradual decline in volcanic activity as the Moon cooled, but it was surely not compatible with random impact strikes unless all the big impactors fell first! However, we now have a much more refined understanding of the history of our solar system, and it is clear that the big asteroids and comets *did* indeed fall first, during the Late Heavy Bombardment. Since that era both the incidence of such impacts and the size of the remaining impactors have gradually declined to the level of the

89

small-scale and infrequent impacts we see today.

In addition to the grand alignments of large craters discussed above, many smaller crater chains may be found dotted about the lunar surface, looking for all the world like strings of pearls. These, too, were regarded by volcanists as clear evidence that craters formed from within along surface cracks, for surely meteors and asteroids did not hunt in packs! Yet, once again, subsequent events have shown that such features are entirely compatible with the behaviour of impactors. The approach of Comet Shoemaker–Levy 9 to Jupiter in 1994 showed how a relatively small body can be torn apart by the gravitational forces generated by a larger one. As the resulting fragments of Shoemaker–Levy 9 ripped into the upper layers of the giant gas planet and burned up, they left a string of scorch marks that remained visible to telescopic observers for some time afterwards – the atmospheric equivalent of the more permanent impact chains we see on the lunar surface. Also, many of the Moon's crater chains must have been created as secondary impacts made by blocks of material thrown out in aligned arcs during the excavation of a larger crater.

Moore was also struck by the fact that many lunar mountains, including crater central peaks, appeared to possess summit craters, and together with H. P. Wilkins he used some of the largest telescopes in Europe in order to search for examples. For him such pits were reminiscent of the summit calderas found on terrestrial volcanoes, and they were thus further evidence of a volcanic origin for the craters of the Moon. He dismissed out of hand the idea that such

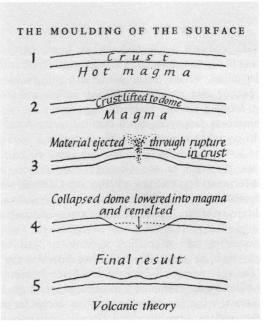

THE MOULDING OF THE SURFACE

1   *Crust*
    Hot magma

2   *Crust lifted to dome*
    Magma

    Material ejected through rupture in crust
3

    Collapsed dome lowered into magma and remelted
4

    Final result
5

Volcanic theory

Volcanic theory of lunar craters, according to Patrick Moore.

pits, perfectly placed at the tops of peaks, could be the result of random 'lucky strikes' by meteorites, insisting that 'the bomb does not fall into the bucket every time!' This view is perfectly reasonable – such coincidences would indeed be infrequent. However, the advent of high-resolution spacecraft imagery has demonstrated conclusively that most of the summit pits detected by Wilkins and Moore are not true pits at all, but an illusion created by inadequate telescopic resolution (despite the large apertures used). The example here shows an enlarged crop from a telescopic image of the central peak in Gassendi. The image is of very high resolution for one taken from Earth and it appears to show two summit pits; but when the same area is modelled using spacecraft data the 'pits' are revealed to be merely an effect created by the 'molar' shape of the central mountain mass.

The 'ghost crater' Lamont, which Fielder described as an elementary ring.

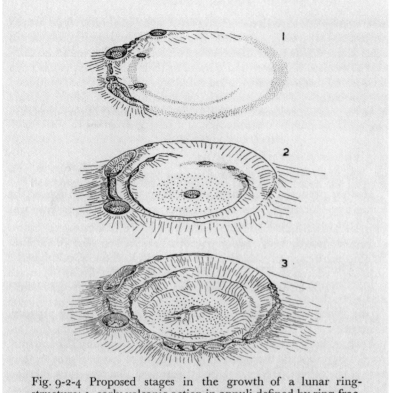

Fig. 9-2-4 Proposed stages in the growth of a lunar ring-structure: 1, early volcanic action in annuli defined by ring-fractures (shaded); 2, complete growth of an outer ringwall, and partial growth of an inner ring, or spiral, by volcanic extrusion. Some subsidence and melting of the floor of the main ring-structure; 3, further in-dropping along ring faults, further development of the outer and inner walls by lava flows from the rim volcanoes, and the replacement of a central orifice by central cones, lineamented in the direction of regional fractures

Lunar ring structure, from Gilbert Fielder's *Lunar Geology* (1965).

The problem with the models proposed by Moore, Cattermole and Fielder was finding a known terrestrial volcanic mechanism that would work at the scales required to create the craters of the Moon. As we have seen, this was a problem that had plagued most earlier advocates of a volcanic origin for such features. Moore fell back on an unlikely variation of Hooke's bubbles, suggesting a sequential uplift and sinking of the lunar crust.

Later, with Cattermole, he favoured a more persuasive mechanism that saw craters like Theophilus begin with eruptions of ash and lava from central vents, which resulted in a central peak complex and summit calderas as well as the withdrawal of magma from an underground chamber. The emptying of this magma chamber led to the collapse of the crater floor and the creation of fractures at what would become the crater's circumference. Later eruptions of ash and lava around these circumferential fractures then gave rise to the system of ramparts.

Fielder advanced a rather different model for at least some lunar craters, based upon a novel interpretation of the so-called ghost rings, those degraded remains of old craters that have been engulfed by mare lavas. Lamont, a system of circular ridges east of the crater Arago on the Mare Tranquillitatis, is a good example and one that Fielder used himself.

Fielder took the revolutionary step of arguing that craters such as Lamont were not at the end of their lives, but at the very start. They were elementary rings in the process of emergence as a result of lavas extruded around ring faults. Subsequent 'in-dropping' within ring faults would give rise to the fully formed crater.

The work of Fielder, Moore and Cattermole represented a significant rearguard action against the advance of impact theory, but as the 1960s moved on their efforts seemed increasingly like whistling in the wind. It was only a matter of time before the end for volcanic theories of formation for the majority of the Moon's craters. This end finally came with the Apollo lunar landings, and the retrieval of rocks that had been shattered and melted by shocks that could be commensurate only with high-energy impacts.

FOUR

The Moon in the Age of Spacecraft Exploration

Within a half-century or so of the dawn of the Space Age in October 1957, spacecraft succeeded in providing answers to many of the questions asked about our Moon not just since the invention of the telescope, but over the centuries of observation and speculation preceding it. Certainly there is yet more to learn, and each answer seems to beg further questions, but the Moon is no longer the terra incognita that puzzled our ancestors. We are not able to consider in any detail here the individual results achieved by those missions – manned and unmanned, Russian and American, Chinese, Indian and Japanese – that have added to our fund of knowledge. In any case, such a task has already been admirably performed elsewhere.[1] In this chapter we must limit ourselves to a review of lunar science as we now understand it thanks to the results of those missions.

The first truly significant achievement in lunar exploration by spacecraft was Luna 3, the Soviet mission to photograph the Moon's far side, in October 1959, only two years after the dawn of the Space Age. As we have seen, the Moon's captured rotation, synchronous with its period of revolution around the Earth, means that to all intents and purposes it keeps one face permanently turned towards us and the other permanently hidden from the Earth-based observer. But, as Arlin Crotts nicely says, 'nature may not abhor a vacuum, but the human mind does,' and before Luna 3 many had speculated about the likely nature of the Moon's mysterious hidden hemisphere.[2] That averted hemisphere is often erroneously referred to as the Moon's 'dark side', when in fact it receives just as much light as the near side. But its very inaccessibility has long enhanced its romantic allure.

Most had anticipated that it would be similar in appearance to the visible hemisphere, with bright, heavily cratered upland areas and smoother, darker patches of maria. After all, it was reasonable to assume that the Moon was a consistent body and uniform over its entire surface. Some, however, had argued that the Moon's averted hemisphere might indeed live up to its mysteriousness and turn out to be entirely different.

Perhaps the most remarkable example of such thinking came from the Danish-German mathematician and astronomer Peter Andreas Hansen (1795–1874). In a paper presented to the Royal Astronomical Society in 1854 (published in 1856) Hansen ascribed perceived anomalies in the Moon's motion to irregularities in its figure and mass concentration. He concluded that the Moon was not perfectly spherical but egg-shaped, with its centre of gravity offset from its centre of figure by some 59 km (37 miles) in a direction away from the Earth. The lunar surface on the Earth-facing side was thus higher with regard to the Moon's centre of gravity than that on the averted hemisphere. According to Hansen, that eccentricity would be enough to encourage any lunar atmosphere and water to pool in the lower-lying areas of the Moon's far side, thus allowing the possibility of life in those areas, while the higher near side remained as sterile, bleak and lifeless as we see it.[3]

Hansen's fanciful hypothesis did not stand the test of time, even though it was popularly referenced in Jules Verne's novels *From the Earth to the Moon* (1865) and *Around the Moon* (1869). Others, including some later amateur observers, adopted a less imaginative way of envisaging the Moon's hidden hemisphere. The twentieth-century British observers H. P. Wilkins and E. F. Emley tried to

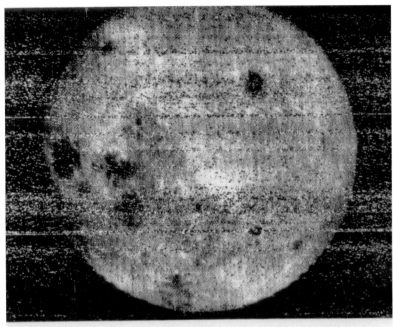

Φοτογραφия 1

Photograph of lunar far side returned by Luna 3 in October 1959. The near-side Mare Crisium is identifiable at left, but the averted hemisphere shows few sizeable mare patches.

trace back bright rays coming over the lunar limb from the far side in order to locate possible ray craters similar to Tycho. But the true nature of the lunar far side remained unknown until the first images were returned from Luna 3. Some 29 images were taken on 7 October 1959, covering 70 per cent of the hidden hemisphere. These were processed on board and then transmitted as the spacecraft approached Earth on 18 October. They were of poor quality by today's standards, but good enough for the subsequent creation of a map.

They were also of sufficient resolution to show that although the averted hemisphere was similar to the lunar near side in that it was pockmarked with craters, it was significantly different in its lack of sizeable dark maria. Apart from a couple of smallish dark mare patches (the most obvious of which was immediately named the 'Sea of Moscow' (Mare Moscoviense), the rest of the newly

The near-side Moon before and after the maria lava flows filled pre-existing impact basins.

uncovered hemisphere seemed to consist of bright highland terrain. Indeed, whereas about 30 per cent of the hemisphere visible from Earth is covered by maria, only about 2 per cent of the far side is. Moreover, the near-side maria can be huge (Mare Imbrium is over 1,000 km (621 miles) in diameter), but the few far-side ones are tiny in comparison.

So, while the longstanding question of what the far side looked like was answered by Luna 3, another – just as puzzling – was raised: why are the two hemispheres so different? As we saw in the previous chapter, the lunar maria were formed when molten magma generated by the heat of radioactive decay and the subsequent melting of rock in the Moon's mantle erupted on to the lunar surface over 3 billion years ago, filling the floors of impact basins that had themselves been created some hundreds of millions of years before that. The crustal fractures created by the violent formation of those basins provided the conduit for the magmas to reach the surface. The lower density and lower viscosity of the lunar lavas, aided by the Moon's weaker gravity, allowed the surface lava flows to spread widely enough to fill even the largest impact basins. What is puzzling is that impact basins are just as prevalent on the lunar far side, and they can be just as sizeable as those on the Earth-facing hemisphere. Indeed, the South Pole–Aitken Basin, which is located largely on the averted hemisphere, is 2,500 km (1,550 miles) in diameter, the biggest impact basin on the Moon and one of the largest in the solar system. It is just that the majority of the far-side basins have not been subjected to the subsequent lava flows and infilling that have created the dark near-side maria.

It is not clear even now why this should be so. It has been suggested that the Moon's crust on the Earth-facing hemisphere might be more susceptible to disruption by the Earth's gravitational influence, giving rise to a greater incidence of crustal fracturing, which would in turn permit the easier outflow of molten lavas. Experiments conducted during the Apollo missions, along with data collected from later automated spacecraft, have revealed that the Moon's crust is indeed thinner on the Earth-facing side and that the Moon's centre of mass is displaced by a small amount (approximately 2 km (1.2 miles)) towards the Earth – quite the opposite of what Hansen once argued! We do not know why this should be so, although it has been suggested that the far-side bulge might have been the result of a huge asteroidal impact in the Moon's early history, one that created a 'megabasin' on the near side. A huge amount of material might have been ejected into space by such an event, but much of it at less than escape velocity, so that it was carried around to the Moon's far side and deposited

False-colour elevation models showing the relative surface heights of the Moon's near side (left) and far side (right). Blue colours indicate low-lying areas and reds higher terrain. The South Pole–Aitken Basin is clearly picked out in blue at the bottom of the far-side image, but the majority of that hemisphere is dominated by highland terrain.

at the antipode. This would give rise to the gravitational offset we observe and the imbalance in crustal thickness between the Moon's two hemispheres.[4] However it came about, might such an imbalance be enough to explain the preferential pooling of lavas on the Earth-facing side? For if we assume that molten magma is of the same density and viscosity throughout the Moon and that it rose from the same depth, it would not have had as far to rise in order to flow out onto the surface of that side.

A recent article by Paul Spudis has argued against this explanation, since it would appear that there are significant localized variations in the density of the lunar mare lavas. As Spudis writes:

> That implies that, even *if* material all came from the same depth (unlikely), it wouldn't necessarily have risen the same distance. Thus, the contrast in the number of near- and farside maria can't merely be the result of magmas of similar densities rising to similar levels.[5]

Spudis also considers the suggestion that differences between the two lunar hemispheres might be attributable to the uneven distribution of radioactive elements, leading to a greater concentration of heat-producing minerals and the eruption of more molten lavas on the near side. But this only pushes the problem back a stage, for why should the radioactive elements have been unevenly distributed between the near and far sides in the first place? In the end, as Spudis concedes, we do not yet know why our Moon should be two-faced, and the mysteries uncovered by Luna 3 still puzzle us more than half a century later.

But if the flotilla of manned and unmanned space missions that have visited or orbited the Moon since Luna 3 have yet to resolve the enigma of the Moon's different hemispheres, they have done much to uncover the true nature of our satellite's formation and subsequent geological history. It is now possible to advance with some confidence a reasonable account of the Moon's development from the time it was brought into being following the 'Giant Impact'. That great collision between the proto-Earth and a Mars-sized planetesimal (Theia) remains hypothetical and leaves many questions unanswered, but it is still the most favoured theory of how the Moon came to be formed. We saw in Chapter One how, by the time the Giant Impact occurred some 50 million years after the birth of the solar system, Earth and Theia, although semi-molten, had been around for long enough for both bodies to have undergone a process of differentiation. That process led to the heavier metallic elements (siderophiles) separating out and sinking to form an iron-rich core, whereas lighter minerals (silicates) rose and floated to form a mantle and crust. It was that lighter mantle material that was thrown off both bodies by the impact, forming a debris disc from which our Moon later coalesced. This also explains why the Moon is significantly (about 40 per cent) less dense than Earth, as the latter absorbed what was left of Theia, including its iron-rich core, leaving the Moon depleted in siderophiles and with only a relatively small metallic core, less than 350 km (217 miles) in radius and comprising only some 2–4 per cent of the total lunar mass. Seismographic data from Apollo missions, along with data from orbiting spacecraft, has confirmed the relatively small size of the Moon's core, which appears to consist of a solid inner metallic core (160 km (99 miles) radius) surrounded by a fluid outer metallic core.

The nascent Moon was hot, melted by the heat of the Giant Impact as well as by gravitational energy as it accreted, and this gave rise to a deep, planet-wide magma ocean. That same heat also drove off volatiles, which explains the low abundance of water locked up in the lunar rocks returned by Apollo. As the Moon accreted it underwent further differentiation, as minerals solidified from the magma ocean and either sank or rose, depending upon their density. Heavier minerals such as olivine and pyroxene precipitated out of the magma ocean and sank to form the deep lunar mantle. Lighter minerals, such as aluminium-rich anorthosites, rose and floated in a scum-like layer above the mantle that then cooled and solidified to form the original lunar crust. Those anorthosites make up the lighter lunar highlands and they are the oldest rocks on the Moon. One sample, brought back by the Apollo 15 astronauts in 1971, is known as the 'Genesis Rock'. It has been dated as over 4 billion years old and is now held as Sample 15415 at the Lunar Sample Laboratory Facility in Houston, Texas.

However, some minerals, known as 'incompatible elements', did not combine easily into the rock structures that separated out from the cooling magma, and these remained in what was left of the liquid magma ocean, sandwiched between the anorthositic crust and the olivine/pyroxene-rich mantle. These isolated elements included potassium (K), various Rare Earth Elements (REE) and phosphorus (P), giving rise to the wonderful acronym KREEP. Movement, churning and turnover in the remaining magma ocean, along with heat generated by further impacts and from the decay of radioactive elements in the KREEP layer, kept the lunar mantle partly melted even as the Moon cooled. It was this molten magma that later poured out onto the lunar surface, filling the giant impact basins and creating the extensive lava plains of the lunar maria. It also contributed to other forms of lunar volcanism, which we shall consider in due course. Analysis of the Moon's surface chemistry by spacecraft has confirmed that KREEP material is largely associated with the large maria and lava flows of the Moon's near side, especially in the Oceanus Procellarum.

The relatively small size and low mass of the Moon meant that it cooled more rapidly than the Earth. This cooling, combined with the smallness of its metal core, resulted in the Moon being unable to retain a liquid metal core of sufficient size to generate a global magnetic field such as we find on Earth. However, it does show evidence of localized magnetic fields often associated with mysterious bright swirls scattered around the lunar surface.

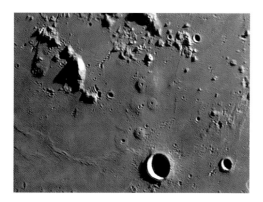

A field of volcanic domes near Hortensius.

But the dark features of the man in the Moon's face are the results of subsequent volcanic activity and were formed when molten basaltic lavas erupted onto the surface typically between 3.5 and 2.5 billion years ago (although some are younger). They cover some 7 million km^2 of the lunar surface – ample evidence in itself of the importance of volcanism in sculpting the Moon's topography. That the 'seas' (maria) are more recent features than the highlands is betrayed not only by their different composition, but by differences in the density of impact cratering. The ancient highlands have been battered by impacts over a long period of time and they are crowded with layers of craters of various ages, the more recent overlying and degrading the older ones. The maria, on the other hand, show evidence only of those impacts that have occurred since their formation, with the result that their surfaces are relatively smooth in appearance.

We saw earlier that the Moon's lavas were less viscous than terrestrial lavas and its gravity much less than Earth's, with the result that lunar lava was able to flow extensively and produce maria of great size. Even one of the smallest, Mare Crisium, is over 500 km (310 miles) across at its widest point. The larger Mare Imbrium is some 1,100 km (683 miles) in diameter, while the lava flows making up the less well-defined Oceanus Procellarum cover an area of around 4 million km^2. However, not all of the Moon's volcanic processes were on such a huge scale. Small volcanic domes averaging about 10 km (62 miles) in diameter are common, mainly on the maria and the floors of lava-filled craters. These have very gentle slopes, rise to a height of only a few hundred metres, and are visible only under glancing illumination. They often have summit pits, which are the sources from which lava flows erupted

109

onto the adjacent surfaces. The fact that these lavas cooled locally to form small-scale volcanic domes suggests that they may have been denser, cooler and less fluid than those that had filled the extensive impact basins and created the maria. As well as through the extrusion of lavas onto the surface, domes may also have formed as a result of the intrusion of magma laterally into subsurface zones of weakness, causing the surface layers of rock to lift and bulge. Lunar domes formed by extrusion are similar to terrestrial shield volcanoes, while those that formed as a result of subsurface magma intrusion are akin to terrestrial laccoliths.

The processes of lava extrusion and subsurface magma intrusion also gave rise to other volcanic landforms on the lunar surface. Extruded lavas often created winding, river-like channels as they flowed from volcanic vents. Sometimes such channels became roofed over as the surface lavas cooled and hardened while the molten materials continued to flow and drain underneath. In places those roofs later collapsed, exposing the lava tubes beneath. These features are known as sinuous rilles, and the best examples are Schröter's Valley, which flows from a large vent known as the Cobra

Two examples of sinuous rilles. Left: Schröter's Valley; right: Hadley Rille.

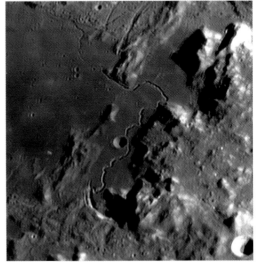

The Ariadaeus Rille.

Head on the Aristarchus Plateau, and the Hadley Rille, which was visited by the Apollo 15 astronauts in 1971. Such lava tubes are also to be found on Earth in areas of extensive volcanism, such as Iceland.

But, as with domes, rilles can also be the result of intrusive volcanic processes. If magma is forced into vertical dykes in the lunar crust it can give rise to faulting on the surface. This in turn can create linear rilles, which are similar to the graben we see on Earth and are formed when the surface slumps between parallel faults. There are many such examples on the Moon, but one of the most spectacular is the Rima Ariadaeus, which lies to the south of the large ruined crater Julius Caesar. A third type of rille, although one created only indirectly through volcanic processes, is the arcuate rille; we see these concentric to the 'shores' of some lunar maria. These are essentially cracks created by the weight of the lavas that filled the impact basins, which caused slumping of the basin floors and tensional faulting at the edges.

The features described so far – the maria, domes and rilles – have been the result of relatively quiescent forms of volcanism: the extrusion or intrusion of magma and the flow of lavas over the lunar surface. But the Moon also shows evidence of a more explosive type of volcanism in the form of pyroclastic deposits, that is, deposits of ashes or glasses that can only have been formed through violently eruptive processes. Here we have to acknowledge that some of the craters of the Moon are indeed of volcanic origin. However, they are relatively few in number and small in scale when compared to impact craters. It is in their vicinity that we tend to find evidence of pyroclastic volcanism. For example, observers have long noted the existence of dark, roughly circular patches surrounding small crater

Volcanic vents and dark pyroclastic ash deposits in Alphonsus.

pits within the large impact crater Alphonsus, and those features have been termed 'dark-haloed craters'. Apart from their small size, they are different in other ways from their impact brethren. Most notably, they lack the raised rims of excavated material found in impact craters, and they lie along surface fractures and rilles. These crater pits are volcanic vents, and the dark haloes associated with them are deposits of ash erupted under great pressure and at very high temperatures.

The most extensive area of pyroclastic deposits on the Moon is the Aristarchus Plateau, mentioned earlier. One does not have to look far to identify their source, for the Cobra Head at the start of Schröter's Valley is one of the largest volcanic vents on the Moon. At Full Moon the keen eye at the telescope should just be able to

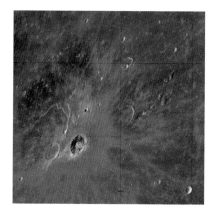

The Aristarchus Plateau. Clementine colour-ratio data showing the different mineralogical composition of the plateau and the surrounding mare.

The Hyginus Rille is dominated by small collapse pits, especially along its western arm. Note the delicate rilles connecting the Hyginus Rille to the Triesnecker system to the south and the Ariadaeus Rille to the east.

discern that the entire plateau has a faint yellowish hue that clearly marks it out from the surrounding grey mare lavas that abut it to the north, west and south. This suggests that it is blanketed in different materials. It also implies that the pyroclastic materials that cover the plateau were erupted before the lavas that surround it, otherwise they, too, would be draped in the same stuff. The entire area of the Aristarchus Plateau is a volcanologist's paradise. Apart from possessing the largest volcanic vent on the Moon and the largest sinuous rille, it shows further evidence of a volcanic past to the east, in the direction of the ruined crater Prinz and the Harbinger Mountains. Prinz is adjacent to an elaborate collection of sinuous rilles, smaller than Schröter's Valley to be sure, but impressive lava tubes nevertheless. Moreover, that area to the east is draped in what looks like the same pyroclastic material as the Aristarchus Plateau itself.

Another area that has been largely sculpted by volcanic activity is that around the small 10-km (6-mile) crater Hyginus, located at the junction of the Sinus Medii and the Mare Vaporum. Rilles dominate the locality: the Ariadaeus Rille lies immediately to the east, the Triesnecker system of fractures lies to the south (probably caused by magma intrusion and crustal uplift) and Hyginus itself sits astride an unusual rille that appears to be connected to the Ariadaeus and Triesnecker systems. It is immediately obvious that the Hyginus Rille is not a sinuous rille like Schröter's Valley; nor is it like the Ariadaeus and Triesnecker linear rilles. It is in two parts that are aligned at an angle on either side of Hyginus itself. Moreover, it is distinguished by chains of small crater pits that intrude upon parts of its length. These pits are rimless, so they are not impact features. They are almost certainly volcanic collapse

pits, or calderas, and they perhaps formed along lines of crustal
fracture coincident with the alignment of the Hyginus Rille.

Hyginus itself is also rimless and probably volcanic in origin
– simply the largest of the collapse pits along the length of the
Hyginus Rille. It is flat-floored rather than bowl-shaped like most
small impact craters, and its floor contains many small dome-like
mounds that are only just visible telescopically, but which turn out
to be much more interesting when seen from spacecraft. Close-up
imagery returned from the Lunar Reconnaissance Orbiter (LRO)
reveals these mounds to be surprisingly similar to those found in
an unusual lunar landform that was first identified in Apollo 15
imagery. Originally referred to as the D-feature because of its shape,
but now called Ina, it struck its discoverers as being unlike anything
previously seen on the lunar surface. Located on the Lacus Felicitatis,
a small smudge of mare material southeast of the crater Conon, Ina
looks like a caldera, a maximum of 3 km (1.9 miles) in width and
about 60 m deep (197 ft), located at the top of what appears to be a
domed uplift. It is a very difficult object for the telescopic observer,
but LRO imagery suggests that it is a shallow depression with a floor
that is dotted with strange smooth, elevated patches that look for all
the world like pooled drops of liquid. These appear similar in nature

Hyginus (left) and Ina
(right). Note the smooth
bead-like mounds on the
floors of both.

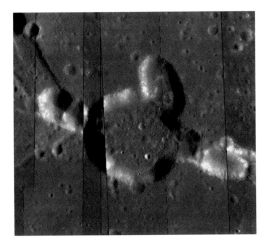
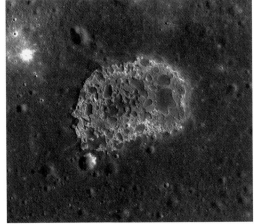

to the surrounding mare material and they are pitted with a similar density of small impact craters, but the lower ground between them is quite different, being extremely rough in texture and with fewer impact craters. This suggests that the lower parts of Ina are younger, with some researchers suggesting an age measured in millions of years rather than the billions of the surrounding mare.

At first Ina was thought to be unique, but LRO imagery has subsequently revealed the existence of some seventy similar features. These are now known as Irregular Mare Patches, or IMPs, but their true nature remains uncertain. There appear to be two contending explanations. One is that the smooth beads of mare material represent young lava flows, something that would have serious implications for our current understanding of the timescale over which the Moon's cooling occurred. The second suggestion is that the rough lower areas have been excavated by some process of outgassing that removed the regolith of mare material within features such as Ina, leaving only patches of it in the form of the beads we see. The debate goes on, but what seems incontrovertible is that both explanations rely upon volcanic processes that appear to have been active in the relatively recent geological past.

So where does this leave us? Certainly the Moon remains largely dead, a museum displaying the history of our solar system's violent past. But the results returned by each generation of lunar orbiters and robotic explorers provide fresh information about our satellite, and this continues to confound our expectations and force us to reappraise what we thought we knew. We can be confident that we understand pretty well the processes that gave rise to the Moon's impact craters and basins; but the discovery of new landforms like Ina suggests that we still have much to learn about our volcanic Moon.

Until quite recently the generally held view was that the Moon was a dry and arid world, any water originally locked up in the materials from which it formed having been driven off by the immense temperatures generated during the Giant Impact. This view reflected a wider belief that water itself, in all its forms, was a rare commodity in our solar system. Those who departed from this consensus were often dismissed as cranks and wishful thinkers. We have already seen how the ideas of Hansen, suggesting that water and air might have gathered on the far side of a Moon in which mass was unevenly distributed, were quickly dismissed. The suspicion that such ideas might be largely the preserve of the eccentric was reinforced by the 'Glacial Cosmogony' (Glazial-Kosmogonie) of the Austrian engineer–inventor and amateur astronomer Hanns Hörbiger (1860–1931), who argued that water, in its 'cosmic' form of ice, dominated the universe in a dialectical struggle with its antithesis, fire. Hörbiger's theory – developed in collaboration with the German schoolteacher and selenographer Philipp Fauth, who also believed in an ice-covered Moon – came to him not as a consequence of scientific deduction, but in the form of a semi-mystical vision. It was nonsense of the first magnitude, and it is hardly surprising that (now with the newly coined Germanic name Welteislehre – 'world ice theory') it was favoured by the Nazis both before and after Hitler came to power, its 'visionary' nature and Austrian origins a perfect antidote to the 'Jewish science' of Albert Einstein.

However, times have changed, and thanks to advances in space exploration we now know that water is to be found almost everywhere in the solar system, locked up in the materials making up asteroids and comets, frozen in immense layers of water ice just below the surface of Mars, spouting from geysers on Saturn's moon Enceladus, and probably forming extensive subsurface oceans on planetary satellites such as Europa, Ganymede, Enceladus and Titan. Even Mercury, the smallest planet and the one closest to the Sun, with a blistering daytime surface temperature in excess of 400°C (752°F) hides deposits of water ice in permanently shadowed craters near its poles. Extensive deposits of water ice have also been found on Pluto, thanks to the remarkable results returned from the New Horizons mission. It is therefore not surprising that the consensus of opinion now favours the presence of water on our Moon too. Of course, the idea held by observers of the past that expanses of water were to be found on the lunar surface and that the maria were indeed seas, remains unsupportable, as it has been since it was recognized that the Moon has no appreciable atmosphere, that its daytime temperatures are extremely high and that any exposed expanses of water or ice would quickly sublimate away.

However, many clung to the idea that surface water might have existed in the Moon's past: the wrinkle ridges found on mare surfaces were often explained as the results of alluvial action on the beds of ancient seas, and sinuous rilles were regarded as dried-up river beds left by turbulent water flows. Similar ideas were still being defended as late as the 1960s by some scientists, including the Nobel Laureate Harold Urey, who interpreted the Ranger 7 close-ups of crevasses of the floor of Alphonsus as probable evidence of water having flowed and evaporated from below the surface.[8] Initial studies of rock samples returned by the Apollo missions threw up no evidence of water present in the lunar rocks, and the sinuous rilles (such as the Hadley Rille, visited by the Apollo 15 astronauts) were confirmed as sites where lavas, rather than water, had once flowed. This absence of water was entirely in keeping with the giant impact theory of the Moon's formation, in which volatiles such as water would surely have been driven off in the high temperatures generated by the collision.

Since then the picture has changed dramatically. Later analysis of Apollo samples indicated some evidence of hydration, although this was initially put down to probable contamination of the samples by terrestrial intervention. Subsequent robotic missions to the Moon have, however, provided more compelling evidence of the presence of water. Some of this evidence, such as the possible detection by the Clementine mission (1994) of water molecules in the lunar regolith, was ambiguous. However, more recent probes such as Chandrayaan-1 and the LRO have found what look like water ice deposits on the floors of permanently shaded craters near the Moon's poles, and the evidence here seems much more secure and entirely consistent with the detection of water in equally unpromising locations elsewhere in the solar system. Moreover, NASA's *Moon Mineralogy Mapper* (M3) instrumentation, carried on the Indian probe Chandrayaan-1 in 2008, also detected evidence of hydration in the lunar regolith as a whole, and not just at the shaded polar areas where the Sun never penetrates. Even more decisive evidence of water ice deposits was provided by the LCROSS (Lunar Crater Observation and Sensing Satellite) probe carried along with the LRO. LCROSS was deliberately slammed into the crater Cabeus near the Moon's south pole on 9 October 2009, and the plume of material excavated by its impact was analysed by LRO. The result was the detection of significant amounts of water, 'not by the glass but by multiple barrels full', according to Crotts.[9]

So, where did all this water come from? Is it magmatic water originating from within the Moon itself, or was it delivered by external agents such as water-bearing comets and asteroids or the effects of the solar wind when it interacts with the lunar regolith? Or is it a result of a combination of such processes? The simple answer is that we do not yet know, and we are unlikely to make significant progress towards a definitive answer without further sampling. It is clear, though, that if the Moon's water does come from within, we would have to rethink certain aspects of the giant impact theory – in particular the notion that all volatiles were driven off during that event. Admittedly, the detection of water molecules locked up in lunar rocks and of pockets of water ice in the frozen interiors of permanently shadowed polar craters is not in the same league as the discovery of subsurface oceans or extensive sheets of ice elsewhere in our solar system; but it does alert us to the fact that our Moon is not the totally dry world we once imagined. So, if the

Moon is not desiccated, might it also not be entirely dead and devoid of change?

Although it is a given that large-scale geological activity on the Moon ceased some millions of years ago, reports of small-scale physical changes (as opposed to apparent changes caused by variations in lighting and libration conditions) go back deep into the historical record – something that was considered in part in Chapter Two. The earliest such report goes back perhaps as far as AD 557, while better known is the observation made in 1178 by five monks from Canterbury who claimed to have seen a dramatic, fiery flash of light, complete with 'hot coals and sparks', near the northern cusp of the young crescent Moon. The monks were the only witnesses and, if they saw anything at all, it is likely to have been an event in the Earth's atmosphere, such as an exploding bolide approaching from a direction coinciding with the Moon's position in the sky. More recently, we might recall William Herschel's observation of 'glowing volcanoes' on the Moon's unlit hemisphere in April 1787 – almost certainly bright ray craters lit up by earthshine – or Julius Schmidt's famous observation in October 1866 of changes in the appearance of Linné.

The most significant modern reports of lunar changes, or 'transient lunar phenomena' (TLP), include Nikolai Kozyrev's

Lunar Horizon Glows imaged by the Surveyor 7 lunar lander in 1968.

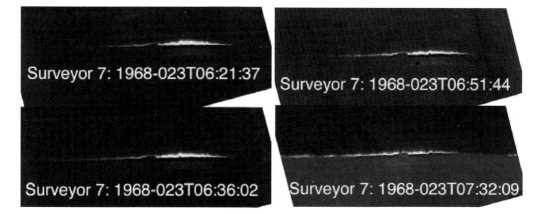

Surveyor 7: 1968-023T06:21:37

Surveyor 7: 1968-023T06:51:44

Surveyor 7: 1968-023T06:36:02

Surveyor 7: 1968-023T07:32:09

claims to have recorded gaseous emissions in the crater Alphonsus on 2 November 1958; the reports of red glows in and around Aristarchus by James Greenacre and Edward Barr of the Lowell Observatory, Flagstaff, Arizona, on 29 October 1963; and Audouin Dollfus' observation of anomalous glows in the crater Langrenus on 30 December 1992, using the 99-cm (39-in.) reflector at Meudon, Paris. Moreover, astronauts on the Apollo missions reported occasional glows and bright flashes while they were in the vicinity of the Moon. The flashes were quite possibly impact flashes, whereas the surface glows, reported in Aristarchus by the crew of Apollo 11, are rather more doubtful.[10] However, horizon glows had earlier been detected and photographed by NASA's Surveyor spacecraft-lander between 1966 and 1968, as well as by the Soviet vehicle Lunokhod 2 in 1973, and similar horizon plumes of light were seen by astronauts on the Apollo 8, 10, 15 and 17 missions. These could well have been caused by the electrostatic levitation of charged dust particles, or by the photoelectric lofting of regolith material irradiated by the rising Sun. The LADEE (Lunar Atmosphere and Dust Environment Explorer) spacecraft mission of 2013–14 was designed to investigate further the probability of such phenomena, but at the time of writing its results are still inconclusive.

Since records began, more than two thousand reported TLP events have been catalogued, although some of those reports are more robust than others. Many of the reported events seem to occur in the same few places, especially in and around Aristarchus, in the crater Alphonsus and on the floor of the crater Plato. This may or may not be significant, as those areas are popular targets for observers anyway, and there may thus be an element of observational bias involved. The Aristarchus Plateau and Alphonsus are known seats of pyroclastic volcanic activity in the past, while the small craterlets on the floor of Plato have been suspected of variability by generations of observers. The majority of reported

TLP take one or more of the following forms: (a) glows or brightenings; (b) colouration, usually red or blue; (c) obscurations and mists; (d) flashes of light. Many amateur observers turned to TLP hunting in the 1960s when opportunities for meaningful cartographic work diminished with the advent of spacecraft exploration.

The result was an explosion of TLP reports over the next couple of decades, and there can be no doubt that many of those cases were attributable to over-enthusiasm, lack of experience, wishful thinking or observer error. In particular, reports of anomalous colours (especially reds and blues) often failed to take account of the effects of chromatic aberration, produced either by imperfect optics or atmospheric dispersion and resulting in the separation of white light into its component colours. This is especially likely when the Moon is low in the sky and that light has to pass through thicker layers of the Earth's atmosphere. Such effects can easily impart false colour, especially to brighter features. Similarly, many reports of the 'obscuration' of small details – for example, the craterlets on the floor of Plato – were probably the result of variable seeing conditions. The Plato craterlets are elusive at the best of times, slipping in and out of sight as atmospheric turbulence comes and goes.

More persuasive are reports of instantaneous flashes of light attributable to the impact of meteoroids. The frequency of meteors in our own skies is testimony to the fact that similar encounters with interplanetary material must occur quite frequently on the Moon too. Such material, usually ranging in size from grains of dust to small rocks, tends to burn up in the Earth's atmosphere. But the Moon has no such protective layer, and incoming particles, their speed and size undiminished by atmospheric friction, arrive at the surface with their kinetic energy intact. The resultant impact, even of a small body, is enough to produce the release of sufficient energy to give rise to a flash detectable not only from lunar orbit, but telescopically from Earth. Many such impact flashes have been reliably detected, even with the sort of moderate-aperture telescopes to be found in amateur hands. For obvious reasons, they are usually observed on the Moon's unlit hemisphere, and for several years robotic telescopes armed with sensitive video cameras have been employed at NASA's Goddard Space Flight Center to monitor the Moon's night side.

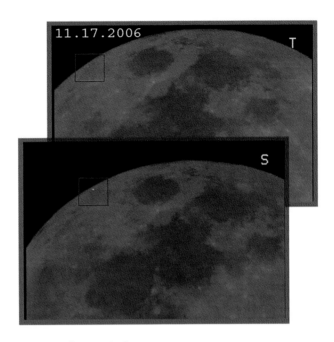

Impact flash recorded by two different telescopes at the Automated Lunar and Meteor Observatory (ALaMO), Goddard Space Flight Center, 17 November 2006.

The video data thus produced can then be fed through detection software. This is observational work that can be carried out successfully by modestly equipped amateurs, some of whom have already participated in pro-am collaborations with the Goddard Space Flight Center.[11]

Recently, Arlin Crotts has made a case for the possible reality of the more robust historical TLP reports. He argues that many reported TLP might be the result of the outgassing of radon-222 from beneath the Moon's surface, adducing as evidence the fact that the presence of this gas has been detected at sites such as Aristarchus, Grimaldi and Kepler, where TLP have been repeatedly reported. The explosive release of such gases, perhaps as a result of moonquakes, would dislodge material from the regolith and blow an expanding cloud of it into space in an event that might be detectable by Earth-based telescopic observers. Mysterious and geologically recent surface features such as Ina and other IMPs could well be the results left by such explosive outgassing.[12] But the jury is still out on the reality of non-impact flash TLP, and

Crotts's views have drawn criticism from the sceptics. What is certain is that, if TLP do exist, they must be quite rare (certainly less frequent than was suggested by the flurry of such reports from amateur observers in the decades following the NASA lunar missions of the 1960s). It will require further missions, both robotic and manned, before the question is resolved.

The study of our Moon during the four hundred years since the invention of the telescope has constituted an important part of the greatest voyage of discovery ever undertaken by mankind: the exploration of our solar system. Initially this was from afar using imagination and ever-improving optics, and then at first hand as our robotic ambassadors have reached out and uncovered the secrets of world after world. It is a story of unmatched grandeur and soaring vision, one involving a struggle against unimaginable distances and immense difficulties, as well as the gradual stripping away of illusions, and the need to balance imagination and understanding. If our species is to be remembered for anything in the eternity to come, it will not be for the petty squabbles that litter our history, but for the nobility of that collective endeavour.

And the voyage is by no means over. As we have seen, the answers provided by

A 35-cm (14-in.) Schmidt-Cassegrain telescope with a focal reducer and astronomical video cameras at the Automated Lunar and Meteor Observatory (ALaMO), Goddard Space Flight Center. This telescope is used to observe the Moon for lunar impact flashes.

telescopes and spacecraft have been outweighed by the new questions
they have raised. Further study from Earth and more interplanetary
exploration will be needed. The process of observation and discovery
begun when humans first asked scientific questions about the Moon
and planets continues today, and there is an unbroken chain
connecting our own efforts with those of Galileo, Hevelius, Herschel,
Schröter and the other great explorers of the past. No doubt great
nations and huge corporations will rise to the challenge of sending
further probes to the Moon and planets, and it is likely that
sometime soon men will once again walk on our satellite. But
there is also a place in this ongoing voyage of discovery for the
humble amateur. Much understanding and personal enrichment
may be gleaned from repeating the observations of the past, and
much can still be done by the dedicated amateur to contribute to
our understanding in the future.

In the next chapter we shall consider practical ways in
which you, too, can participate, even if only in a small way,
in mankind's greatest voyage of discovery.

FIVE

OBSERVING THE MOON

More often than not the Moon is the first object viewed by the beginner newly in possession of a telescope, and it invariably elicits cries of excitement at the wealth of dramatic detail that is revealed even by the smallest of instruments. But that initial enthusiasm often fades, and the Moon tends to be neglected, if not despised, particularly by those who go on to pursue an interest in the observation of deep-sky objects. Its bright light frustrates the efforts of variable star observers and those seeking out the 'faint fuzzies' of nebulae and galaxies lying far beyond the confines of our own solar system. Such observers plan their observational activity for those occasions when the Moon is *not* visible in our sky.

Moreover, since the advent of close-range exploration of our satellite by spacecraft, the idea has gained ground that there is little point in observing the Moon through backyard telescopes. Much clearer views and much more detail may be found on spacecraft

The central peak complex of the crater Tycho as seen from the Lunar Reconnaissance Orbiter. This shows a level of detail that cannot be matched by even the largest Earth-based telescopes.

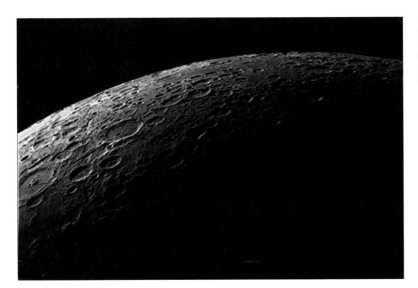

The Moon's southern uplands, showing the level of detail likely to be seen by the inexperienced observer using a small telescope.

imagery, so why suffer the discomforts of sitting outside on bitterly cold winter nights? What can possibly be seen through our telescopes that has not already been seen many times over and better?

There are several answers to these questions. While it is undoubtedly true that the telescopic explorer cannot hope to see the richness of detail revealed by spacecraft imagery, the Moon's constantly changing phases and libration mean that the angles of incident sunlight over its surface are forever shifting, with the result that even familiar features do not look the same from night to night – so there is always something new to see. Some three days or so after New Moon the young lunar crescent becomes evident in the evening sky. Between then and Full Moon the morning terminator sweeps westwards over the lunar surface, each night revealing new features as the Sun rises over them. As these features emerge from the lunar night into daylight they cast long shadows that shorten as the Sun rises higher over them. From Full Moon to New Moon the process is reversed as the progress of the evening terminator sees the shadows lengthen once more, but this time from the opposite direction. Eventually night returns to a given feature about two weeks after it is first revealed.

The cycle is mesmerizing and it repeats itself every month, but not exactly. The exact relative geometry of the Earth, Moon and Sun recurs only over a period of eighteen years, eleven days and eight hours – a period known as the Saros cycle – so it is only after that period that a particular combination of observing conditions can be repeated precisely. There is much for the telescopic observer to gain from close examination of particular areas of the lunar surface over a period of time and under all angles of illumination – not least an understanding of the difficulties faced by observers of the past in interpreting small details and determining the reality or otherwise of reported changes.

Moreover, such sustained examination will provide the observer with not only a sound knowledge of the lunar surface, but valuable training in the art of observation. The latter is by no means easy and it has to be learned. The Moon and planets give up their secrets reluctantly and elusive details on their surfaces will not reveal themselves to the casual observer. The beginner in particular might at first see next to nothing on, say, Mars and might well lose heart. Only with time and experience will the delicate features of the Martian disc be disclosed. The Moon is helpful here for, although it too reveals its finest details only to the practised observer, it also offers immediate encouragement to the newcomer by showing a wealth of features even to the inexperienced gaze. It is thus the perfect place for training the eye to see yet more and for learning the techniques of observation. Nothing matches the immediacy of direct visual observation and the palpable sense of personal participation in that great tradition begun by Harriot and Galileo.

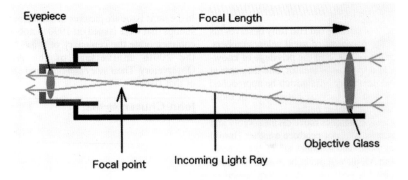

Optical diagram
of a simple refractor.

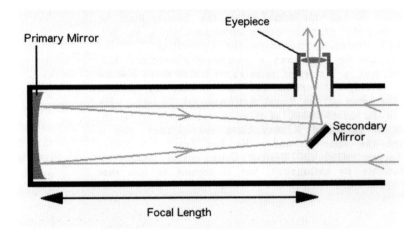

Optical diagram of a
Newtonian reflector.

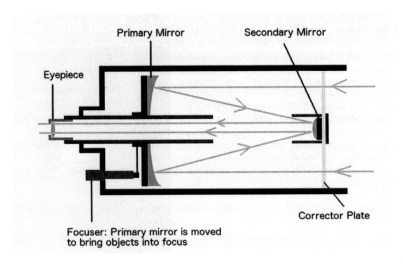

Optical diagram of a
catadioptric Schmidt–
Cassegrain telescope.

Equipment for the Lunar Observer

Although binoculars will provide pleasing views of the Moon's larger features, the observer intent upon proper study will eventually need to acquire a telescope. Nowadays that is much easier than it used to be. There are many established suppliers of astronomical equipment offering a rich choice of quality instruments, often at very reasonable prices. The second-hand market is also buoyant and offers even better value for money. Nevertheless, the purchase of a telescope is not an undertaking to be approached lightly, and the novice should seek advice either from a reputable dealership or from members of a local astronomical society. Generally in astronomy the choice of a telescope comes down to a matter of 'horses for courses', since some telescopes are better suited for particular kinds of observation than others. Once again, the Moon is forgiving in this respect and most proper astronomical telescopes will do an effective job. I am not including here the sort of small instruments often to be found in high-street shops, which are usually overpriced and supported by extravagant claims about their power and magnification – these are quite useless. Genuine suppliers of astronomical telescopes commonly offer three types: refractors, Newtonian reflectors and catadioptics (the last usually of Schmidt–Cassegrain or Maksutov–Cassegrain configuration). This is not the place to go into detail about optical configurations, but, briefly, the refractor produces an image by means of a primary lens, or objective glass, that brings light to a focal point where it can be magnified by an eyepiece.

The Newtonian reflector produces an image by means of a concave primary mirror that is parabolic in figure. Light reflected back from the primary is deflected to the side of the tube by a small flat secondary mirror, and the image produced at the focal point is once again magnified by a suitable eyepiece.

The catadioptric telescope is more complex in configuration in that it uses a simple spherical concave mirror that is easier to produce, but which would introduce aberrations into the image were these not offset by a corrector lens working in combination with the spherical primary mirror. Light reflected from the primary is sent back down the tube by a small convex secondary mirror. It then passes through a hole in the centre of the primary to form an image for the eyepiece.

All three types have their advantages and disadvantages. The refractor is optically stable and rarely needs adjustment, or collimation, of its optical components. It also produces sharp, high-contrast images that are not degraded by the central obstruction supporting the secondary mirrors of the other two types. On the other hand, refractors can produce false colour in their images as a result of the refraction and splitting of white light into its component colours as it passes through the lens. Those components are of different wavelengths and, once separated, they will come to focus at different points, severely degrading the resultant image. This can be partially cured by achromatic objective lenses that use a combination of crown and flint glasses to limit the effects of differential refraction, but such chromatic aberration can only be completely eliminated in high-quality apochromatic refractors, and these are by far the most expensive type of telescope. Anything of greater aperture than 150 mm (6 in.) is likely to be beyond the means of most amateurs.

The Newtonian reflector is much cheaper and, since it uses mirrors rather than lenses, it is largely free of chromatic aberration. However, its optical components require regular and accurate collimation if they are to perform to their capacity, and the presence of a secondary mirror in the light path reduces contrast in the image. Nevertheless, the Newtonian offers the amateur observer the possibility of a large aperture at a bargain basement price, even if the larger the telescope, the more difficult it is to support on a stable mount.

The catadioptric telescope is a 'Jack of all trades' in that it does all things reasonably well but it is a master of none. Its great advantage is its compact size (the result of its folded light path), which makes it easier and cheaper to find a stable mounting, as well as rendering it more comfortable to use. In its Schmidt–Cassegrain (SCT) form it is probably the most commonly used telescope among amateurs today. Like the Newtonian reflector, the SCT needs regular and critical collimation and the large central obstruction of its secondary mirror can reduce contrast in the lunar or planetary image. Maksutov–Cassegrains (MCTs) usually have a smaller central obstruction and are more stable in terms of optical collimation. They can make very fine planetary telescopes.

In the end though, all of these telescope types will provide excellent views of the Moon and are suitable for serious observation. Bigger telescopes will in theory show more detail, but in practice the advantages of greater aperture are often offset by unstable atmospheric conditions, particularly in the UK. For lunar work a Newtonian, SCT or MCT of about 150 mm aperture is a good place to start and will reveal enough lunar detail to keep the observer busy for a lifetime! Refractors of between 100 and 150 mm are also fine, but at that size they are likely to be achromats (unless the owner is rich) and therefore prone to at least some false colour, which will only be enhanced by the Moon's brightness. If an achromatic refractor is your weapon of choice (and many observers are drawn to refractors for their high-contrast images and aesthetic appeal – they do correspond to the popular image of what a 'proper' telescope should look like!), then try to choose one with a relatively long focal ratio of f/10 or more. These will be less prone to the degrading effects of chromatic aberration.

Whatever type of telescope you choose, you will need a range of eyepieces giving low, medium and high magnifications. Three should be enough to start, although ideal magnifications will depend upon the size of telescope and local seeing conditions. For a 150-mm telescope under good seeing, eyepieces giving powers of about ×50, ×120 and ×180 would be ideal. Higher powers might be possible under excellent conditions, but do remember that sharpness of the image is more important than its size. If the image in your eyepiece becomes 'soft', change down to a lower power. Barlow lenses are useful accessories: these are negative, concave lenses that can be used in conjunction with your existing set of eyepieces to amplify the powers available. They are usually sold at ×2 or ×3 amplification, and they are a reasonably cheap way of enlarging the range of magnifications available to you without the need to invest in further eyepieces. The range of eyepieces available nowadays might prove baffling to the beginner. Some cost more than the price of a telescope itself, and these usually provide wide apparent fields of view of 80° or more. The views offered by such eyepieces are indeed spectacular, especially for the observation of extended deep-sky objects and star fields, but they are hardly necessary for close-up observation of the Moon. The beginner in lunar observation would be best advised to invest in a set of moderately priced eyepieces of the Plössl type. These are widely available, quite comfortable to use and generally good value for money.

The Moon can appear uncomfortably bright through a telescope, especially at low magnification. However, unlike the Sun, whose light and heat can cause instant and permanent blindness when concentrated through a telescope or binoculars, the Moon offers no threat to your eyes no matter how bright it might seem. You may, though, wish to reduce its brightness by using an appropriate filter. These come in two varieties: basic neutral density Moon filters that simply cut the glare, and variable polarizing Moon filters that allow you to control the amount of light entering your eye. The latter are probably the better bet. Both types come in a cell with a standard thread that screws into the nosepiece of most conventional eyepieces. An alternative to using a filter is simply to increase the magnification, which spreads the Moon's light over a larger apparent area. Apart from filters designed to reduce glare, there are also specialist coloured filters available, and these can be useful in enhancing contrast between areas of the Moon of slightly different colour and tone. We tend to think of the Moon as monochromatic, but there are examples of subtle localized colour such as the Aristarchus Plateau and the dividing line between the Mare Serenitatis and the Mare Tranquillitatis, where lava flows of different ages and origin meet in a discernible tonal contrast. Experimentation with filters passing different colours can help to bring out such differences.

Apart from optical equipment, the lunar observer will need a reliable guide to the Moon's surface features, and there are various maps and atlases available. The best atlas by far is that compiled by the late Czech selenographer Antonín Rükl.[1] Unfortunately, it is currently out of print, and second-hand copies are much in demand. A useful alternative is the *21st Century Atlas of the Moon* by Charles Wood and Maurice Collins, which uses high-quality imagery from the Lunar Reconnaissance Orbiter (LRO).[2] Do beware of books that have 'atlas' in their titles but which are not true atlases in that they do not cover the entire visible surface of the Moon systematically. These may well be useful books in their own right, but they will prove to be limited as a guide at the telescope.

There are several stand-alone maps available, some produced on laminated paper that will withstand use outside in the cold and damp. It is important that you choose a map with an orientation that matches that seen through your telescope. Binoculars and terrestrial telescopes produce an image that has north up, corresponding to the naked-eye view. Astronomical telescopes usually give an inverted image with south up, but if you use a star diagonal to provide a more comfortable viewing position with a refractor, SCT or MCT, you will end up with an image that has north up, but which is flipped from left to right as in a mirror. Maps exist to match all these configurations, and details may be found in the guide to further reading.

Recording your Observations

As you become more adept at the art of observation, you will almost certainly wish to keep a record of what you see. This is an important development and it is to be encouraged, for it marks the transition from casual Moon gazer to serious observer. All observers should keep a logbook or portfolio containing a record of their observations, including details of date, time, place, telescope, magnification used, seeing conditions and any other relevant data. The two main ways of recording what you see are sketching and imaging, each of which deserves (and has received) much more detailed explanation than is possible here.[3] There is no avoiding the fact that lunar drawing is extremely difficult, and the Moon's complex landforms and constantly shifting illumination provide a stern test for the would-be artist. The important thing is to recognize that accuracy of representation is more important than artistry and that very few have the ability to combine the two effectively. Among those who did were the great nineteenth- and twentieth-century selenographers Thomas Gwyn Elger and Harold Hill, whose work employed a range of techniques in order to depict the intricate details and rich tonalities of the lunar surface. Elger used charcoal or washes of Indian ink, diluted as necessary, to represent the blackest of shadows as well as the intermediate greys of the landscape.

Hill used washes of ink too, but he also developed a stippling technique that came to be much imitated. He would prepare a rough sketch at the telescope, which he liberally annotated with comments and numbers indicating relative intensities, and then he would prepare the final version indoors. Both ink washes and stippling are extremely unforgiving media, and mistakes or spurious detail can easily be introduced unless extreme care is taken.

The reader wishing to learn more about Hill's techniques should seek out his book *A Portfolio of Lunar Drawings*, which is a masterpiece of artistry and accuracy.[4] However, it is not necessary

A page from Thomas Elger's observing logbook for 1887, showing the crater Plato and accompanying observational notes.

134

1887 1st February 5h. 30m to 6h. 30m Definition moderately good

<u>Plato</u>   Power 340.

Shadows on the floor drawn at 6h. 0m

During the whole time of observation I remarked a want of sharpness in the shadows crossing the interior which I do not

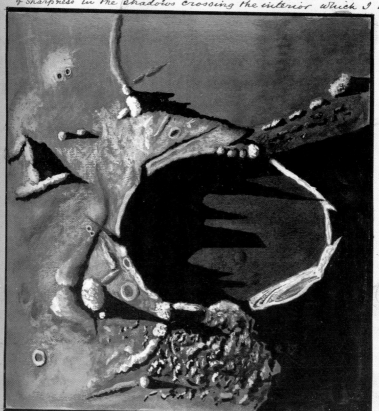

remember having noticed before. The shadows outside the ring and at about the same distance from the ter-minator were on the contrary remarkably clearly defined. The strip of the floor between the shadow of the loftiest peak and the foot of the south-east wall was scarcely distinguishable from the neighbouring shadow. Only one crater-cone was visible.

The crater Moretus and mountains on the Moon's southern limb, captured by Damian Peach using a 350-mm SCT and a high-speed planetary camera on 29 September 2013. Peach is one of the world's finest planetary imagers.

to possess such rare artistic skill in order to produce useful sketches of the Moon, and simple line drawings using pencil can be equally effective as observational records, provided that the observer honestly depicts only what is seen with certainty and care is taken in the accurate placement of features.

Photography offers a real alternative to sketching, and developments in lunar and planetary imaging over the last decade or so have made it possible for the camera to capture much more detail than can be discerned by the eye at the telescope. The displacement of film by digital imaging techniques has led to the production of versatile cameras of great sensitivity. It truly is a golden age for the amateur astrophotographer. Simple photographs of the Moon's entire disc may be obtained using modern DSLR cameras with a suitable long-focus lens, and the same cameras may be used in conjunction with a telescope to secure effective close-up views. Indeed, amazingly detailed images of the Moon

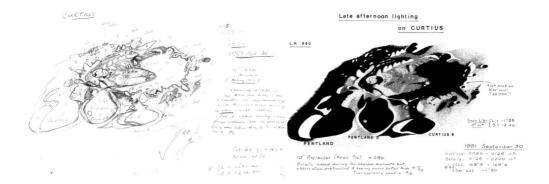

Sketches of the crater Curtius by Harold Hill, showing the rough draft (left) and the finished stippled drawing (right).

may be taken simply by holding up a smartphone camera to the eyepiece of a telescope and focusing the latter carefully!

However, those wishing to obtain really high-resolution images of the lunar surface will seek out a different technique known as 'lucky imaging'. This involves the use of a webcam or specialist high-speed planetary camera, along with a telescope of reasonable aperture and focal length, to secure close-up video captures that can then be processed into highly detailed images. The method involves some investment in equipment and patience, as well as a bit of an initial learning curve, but in essence it is relatively simple and even a less competent photographer can achieve remarkable results. Moreover, the technique allows the observer largely to sidestep one of the greatest difficulties confronting the lunar and planetary observer: the adverse effects of the Earth's turbulent atmosphere. A modern high-speed planetary camera can capture images at up to about 150 frames per second, and this is more than fast enough to capture those fleeting instants of sharp seeing. This means that among the many frames blurred by atmospheric turbulence there will be some that catch those few calmer moments when the detail stands out sharply – hence the name 'lucky imaging'. Those sharper frames can then be automatically extracted from the rest and subsequently aligned, stacked and sharpened using freely available software such as AutoStakkert! and Registax. In the hands of an

137

expert, the technique can produce images that stand comparison with imagery secured from orbiting spacecraft, and it provides the amateur observer with a tool of great power.

An Observer's Review of Lunar Landforms

Once suitably equipped, you are ready to embark upon the telescopic exploration of our satellite. A map or atlas will help you find your way around the Moon's confusing topography, and with practice you will learn to recognize the major features. Some guides recommend that you start your exploration by examining features as they are revealed on the terminator, or day–night boundary, each night. This is fine, but an alternative is to approach the Moon initially by seeking out clear examples of its major types of landform. This will provide a first-hand impression of the Moon's geology, rather than just its geography, and you can in any case start to learn your way around while discovering such examples. In the process you will inevitably confront the effects of the Moon's ever-changing phases. When features are located near the terminator their relief is brought out prominently by the long shadows they cast under a low Sun. Such lighting conditions therefore favour the identification of features such as mountains and valleys, as well as low-relief structures such as domes and

Older lunar craters. Fracastorius (left) has had its northern wall breached and its floor covered by later lava flows. The floor of John Herschel (right) is draped in later ejecta deposits, probably from the impact that created the Imbrium basin.

ridges, which may not be detectable at all under higher illumination. High solar lighting, such as is to be found around Full Moon, will, however, aid in recognition of features displaying colour and albedo differences. Examples of the latter include bright rays and swirls, which can be much more difficult, if not impossible, to detect under low illumination.

Craters

The first thing to strike the newcomer will be the sheer abundance of craters. They are everywhere, ranging in size from great walled plains exceeding 150 km (93 miles) in diameter, down to the smallest pits detectable in any given telescope. You should look out for the differences between simple and complex craters described

Region around the crater Maurolycus, showing complex superimposition of features.

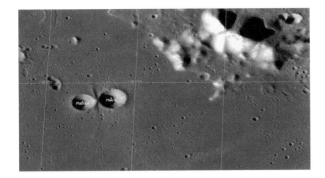

in Chapter Three, with the latter displaying impressive wall-terracing and central peaks (for example, Tycho and Theophilus). Look out also for evidence of craters of different ages. Some older features, such as Deslandres, will have been much degraded by subsequent impacts, while others (for example, Fracastorius or J. Herschel) will have had their floors flooded by later lava flows or ejecta deposits.

Plato K and KA.

Ariadaeus and Ariadaeus A. Note the straight septum between the two, indicating a simultaneous impact.

Younger craters will be recognizable by the sharpness of their features, by the fields of secondary cratering they have created in their vicinity (for example, Copernicus), by their relative brightness, since they have not undergone long periods of space weathering (for example, Aristarchus), or by their bright ejecta rays (Tycho, Kepler). It is instructive to try to work out the relative ages of features in a given area. Apart from the structural differences between old and

The concentric crater Hesiodus A.

fresh craters described above, we can learn much from crater counts and stratigraphic techniques such as consideration of instances where one feature is superimposed on another. As we saw in Chapter Three, younger surfaces will generally carry fewer impact craters, and a feature that imposes upon another will be the younger. Usually such instances are easy enough to decipher, but you will also find craters, particularly in crowded highland areas, where

superimposition of one crater upon another is so complex that it is tricky to reconstruct the sequence of events.

Apart from the broad categories of impact and volcanic, simple and complex, young and old, fresh and degraded, there are also a few unusual variations on the crater theme that deserve special attention from the observer. These tend to be small in size, so they might require some patience and skill on the part of the observer, as well as a telescope of suitable size and favourable lighting and seeing conditions. If you look on the floor of the Mare Imbrium at coordinates 47°N, 3.5°E, some 68 km (42 miles) east of where the Alpine Valley (*Vallis Alpes*) joins Imbrium, you will see a pair of small contiguous craters designated Plato K and Plato KA. They are 6.5 and 5.5 km (4 and 3.4 miles) in diameter respectively, and at the point where they touch there is a discernible low ridge running more or less north–south for some 9 km (5.5 miles) or so.

That ridge, or septum, between the two craters is evidence that they were the result of a simultaneous impact by twin meteoroids and it is possibly a consequence of colliding ejecta blankets. Other symptoms of simultaneous impacts include craters connected by a straight, sharp 'party wall' where they touch. Examples of this may be found in the Ariadaeus/Ariadaeus A pair, near the larger craters Sabine and Ritter on the Mare Tranquillitatis, and the Daly/Apollonius F 'twins' just south of the Mare Crisium. But there are many more instances and the hunt for them would make an instructive observational programme.

Also worth seeking out are examples of concentric craters. These doughnut-like features possess an inner ring that looks very much like a secondary wall. As with the simultaneous impact craters discussed above, they are generally quite small and therefore testing to observe. The most accessible example, and certainly the best known, is the 15-km-diameter crater Hesiodus A, which abuts the southwest wall of the ruined crater Hesiodus on the southern

edge of the Mare Nubium. A telescope of 100–150 mm aperture
should reveal the inner annulus under good conditions.

Such craters are quite rare, with only about fifty known examples
on the Moon's near side. They range from about 2 to 20 km (1 to 12
miles) in diameter, and they tend to cluster on the edges of the lunar
maria. The chances are clearly against the inner annuli being the
results of fortuitous secondary impact strikes within the outer
craters, and we must look elsewhere for an explanation of their
presence. It is possible that the inner rings are the results of debris
slumped from the outer rim, but this seems unlikely in cases such
as Hesiodus A, where the annulus is so regular and smoothly formed.
Are they perhaps the result of a single impactor striking layered
rock beds of different density, giving rise to differential rebound
of materials? The answer is that we don't really know, although
the current consensus of opinion seems to be that the annuli of
concentric craters are volcanic in origin, created either by material
extruded through ring fractures on the floors of pre-existing impact
craters or by the intrusion of magma that lifted the ring. However,
questions still remain unanswered – particularly since there are no
similar features on Earth[5] – and the amateur lunar observer can do
useful work by searching out and charting further examples, perhaps
making use of high-resolution spacecraft imagery freely available on
the web.[6]

Maria

Apart from the abundance and variety of craters, the lunar observer
should also pay attention to the maria, or 'seas'. We have seen how
such maria were formed when molten lavas erupted onto the lunar
surface to fill gigantic impact basins, and evidence of their
comparatively young age is provided by the relative paucity of
subsequent impact cratering on their surface. However, there are
other interesting topographical features on the surfaces of the maria,

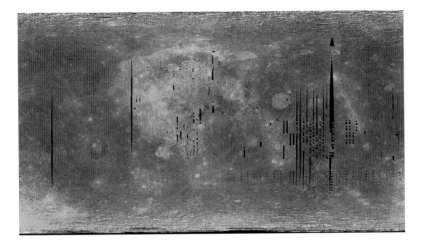

Clementine colour ratio image showing different mineralogical composition of the Serenitatis and Tranquillitatis lavas.

including sometimes striking contrasts in colour and albedo where lava flows of different origins, composition and ages encounter each other. As mentioned earlier, a good example of this may be found on the boundary between the Mare Tranquillitatis and the Mare Serenitatis, where the Tranquillitatis lavas appear darker and bluer in colour imagery.

Colour ratio and mineralogy data from the Clementine spacecraft also show a clear difference between the two, with the central Serenitatis lavas showing up yellow (indicating a lower titanium ratio in their composition) and the Tranquillitatis lavas, along with those at the edges of Serenitatis, showing up blue (indicating a higher titanium content).

Wrinkle Ridges

Also very striking are the wrinkle ridges, or *dorsa* (plural of *dorsum*) to give them their Latin name, which may be found on the lava plains of most of the mare areas, often circumferential to the basins in which they lie. The best-known and most easily observed example is perhaps the so-called 'Serpentine Ridge' near the eastern edge of Mare Serenitatis.

The 'Serpentine Ridge' (Dorsum Smirnov). Note also the albedo differences in the mare lava flows.

Such ridges can run for hundreds of kilometres, although they represent very small differences in elevation and can be easily seen only under low angles of illumination. They are tectonic in origin, caused by the thrusting and buckling of the solidified mare surface as the basin below subsided under the weight of the lavas that had filled it. The result was a series of thrust faults which developed when the fractured slabs of solidified surface lava slid over each other as they were accommodated in the redefined basin space. These faults are relatively shallow in profile, with gentle slopes rarely exceeding a few degrees, and they tend to be discontinuous and made up of smaller aligned ridges more or less concentric with the basin edge. Large mare ridges, such as the Serpentine Ridge, were of course formed several billion years ago, when the basin floors subsided under the layers of mare lavas, but recent imagery from the LRO has revealed the existence of many cognate ridges and scarps that appear to be smaller and much younger. These are known as lobate scarps, and similar features have been found on Mercury. Some of the examples found on the Moon appear to disrupt small craters that are themselves relatively recent in geological terms. This implies that the lobate scarps – and the processes that created them – must be younger still, perhaps occurring less than a billion years ago. It has been suggested that lobate scarps on both the Moon and Mercury are thrust faults that were formed as the parent bodies cooled and their mantles and crusts contracted. This is problematic in the case of the Moon, for it has long been held that our satellite lost its heat quite rapidly and evidence of recent contraction there would suggest that we might need to reconsider quite radically our ideas on how long it took the Moon to cool.

Mountains and Valleys

The observer exploring the Moon's surface for the first time will also be struck by its impressive mountains and valleys, some of which far outstrip in grandeur their equivalents on Earth. For example, Mount Huygens in the lunar Apennines is some 5,400 m (3.4 miles)in height – higher than Everest, which is quite remarkable when you consider that the Moon is a much smaller world than Earth! The Rheita Valley (Vallis Rheita) is about 500 km (310 miles) long, while the Grand Canyon is some 50 km (31 miles) shorter. However, all is not what it seems, and the Moon's mountains and valleys are very different from those on Earth in terms of how they came about. Most terrestrial mountains were created by large-scale shifting and folding of the Earth's crust as a result of plate tectonics. Others were formed by eruptive volcanism. The Moon had no large-scale plate tectonics, and its period of eruptive volcanism was much shorter than Earth's. The Moon's great mountain ranges were created by impact, and they represent what is left of the rims of vast impact basins. Look through your telescope at the lunar Apennine and Alpine ranges and you will see that they form a curve marking out the eastern limits of the Mare Imbrium, where the rim of the

Imbrium basin contained the flow of lavas on its floor. The nearby Haemus mountains form part of the rim of the Serenitatis basin, while the impressive Altai scarp, southwest of the Mare Nectaris, forms part of the outer ring of the Nectaris basin.

Similarly, the Rheita Valley is not a valley of the kind we know on Earth: it, too, is an impact feature – a secondary crater chain

Mountains at the eastern edge of the Crisium impact basin. Note how the mare lavas have flooded and embayed those peaks nearest the basin centre.

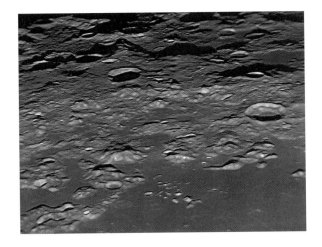

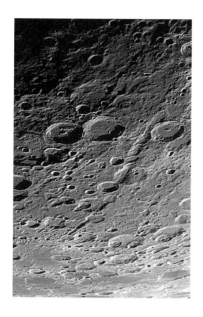

The Rheita Valley.

carved out by huge blocks of material blasted out during the excavation of the nearby Nectaris basin in what must have been an episode of extreme violence.

The Moon's two other best-known valleys have more in common with terrestrial features. As we have seen, Schröter's Valley (Vallis Schröteri) is a lava tube, which once carried molten lava from the caldera at the 'Cobra Head', while the Alpine Valley is a 180 km (112 mile) rift valley, or graben, where the valley floor has slumped between parallel faults in the Moon's crust. At its widest point it is about 19 km (12 miles) across, and a fine sinuous rille runs along its floor for nearly all of its length. This latter feature is visible only in large-aperture telescopes.

Rilles

Rilles were first so called by Schröter after the German word for 'grooves' or 'furrows', but they were an object of fascination for telescopic observers long before then. Also known by the Latin term *rimae* (plural of *rima*), they are found mainly but not exclusively on mare terrain, taking the form of surface channels or fissures that sometimes extend for hundreds of kilometres in length, even though they are only a few kilometres in width. Even a cursory telescopic examination will reveal that there are several different morphological types of rille, and we looked at some of these in Chapter Four. Moreover, these different types were formed by different geological processes. We saw how sinuous rilles, such as the Hadley Rille and Schröter's Valley, correspond to lava channels on Earth, and were carved out when molten lava flowed across the lunar surface from nearby calderas. We also saw how linear rilles, such as the Ariadaeus Rille, are both volcanic and tectonic in origin,

147

caused when magma intruded into parallel dykes and produced faulting and slumping of the surface. Arcuate rilles, also mentioned in Chapter Four, form concentric fissures around the edges of some maria, and they are stress fractures created as a result of the same slumping of mare basins that also gave rise to the thrust faults known as wrinkle ridges. The finest example of such a system of arcuate

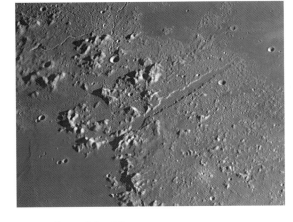

The Alpine Valley. This image also shows the simultaneous impact craters Plato K and KA on the mare surface near the western end of the valley.

rilles is the Hippalus rilles, skirting the western edge of the Mare Humorum. These are some 240 km (149 miles) in length and they are an easy object in even a small telescope when seeing is good and they are near the terminator.

Arcuate rilles are not the only ones that occur in systems. The 26-km (16-mile) crater Triesnecker, located on the Sinus Medii (so-called because it is near the centre of the Moon's Earth-facing hemisphere), is best known for the beautiful and complex system of rilles to its east. These connect with the Hyginus Rille and are in an area of known volcanic activity. They are probably the result of uplift in the crust caused by localized intrusion of magma, and they have fascinated telescopic observers for centuries. A similar system may be found near the 25-km (15-mile) crater Ramsden on the unfortunately named Palus Epidemiarum ('Marsh of Epidemics'). The main rilles of both these systems may be discerned in a small or medium telescope under steady seeing conditions.

Similarly complex systems of rilles may also be found *within* certain types of crater. These are known as floor-fractured craters (FFCs), and they are well worth looking out for as they are formations with an interesting geological history. We saw in Chapter Four how the crater Alphonsus appears to have several pyroclastic vents on its floor. In fact Alphonsus is an FFC and most

The Hippalus arcuate rilles.

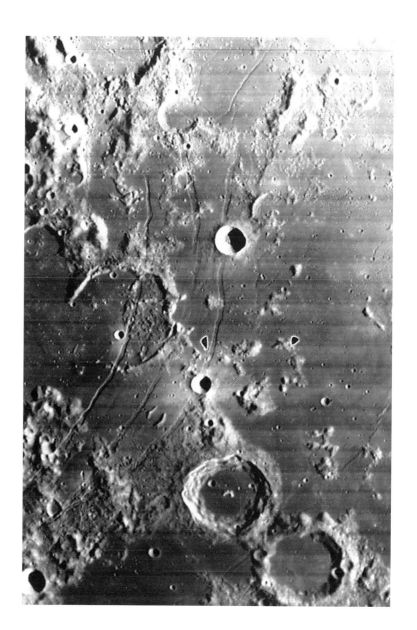

further. The whole of the .... clepts of this system are unusually equal .... in breadth
and depth, with the .... exception
of the ends of .... (1) (2)
and (3) and .... the
two delicate
and rather
irregular
clepts
(4) (5).
The plain
about
Ramsden
is of a uni-
form dark
grey slightly
darker, as is
natural, to the
Eastward.
Ramsden itself
in a bay shut .... is situated
in a bay shut .... in on three
sides by moun- .... tains. On the
S.W. jut out the .... great spurs of the
Capuanus highland and on the S.E. is a curved range of
considerable height. .... containing the
two rings (f) and .... (g) on the S. and
S.E. containing .... a curious depression (j.k), filled .... with shadow.
N.E. of Rams- .... den is a valley
in the centre .... of the open-
ing of which is .... an oval craterlet (n) and another .... object, also, I
think, a craterlet .... (m). Between
Ramsden and (w) .... are several detached
hills in the plain. .... the most prominent
being the two hillocks at the end of Hyppalus 14. (u). (5)
is a long hill with a higher central peak casting a
spire of shade. S.W. of Ramsden and connected with it
by a low neck of hill is a small oval craterlet of the

Ramsden

Nov. 29th 1892.

10.0 L.T. (375).

A. 4.12.92

KEY PLAN.
M. 4.12.92

A page from the
observational notebooks
of P. B. Molesworth,
showing the Ramsden
rille system as sketched
on 29 November 1892.
South is up in this
telescopic observation.

of its volcanic vents are arranged along the rilles that fracture its floor. Further easily observed examples of FFCs include Posidonius, Gassendi and Petavius, while the 207-km (129-mile) Humboldt, although much more difficult to see because of its proximity to the southeast limb, is a spectacular example when the libration is favourable.

Apollo 15 provided a splendid high-resolution image of Humboldt showing that the floor of this crater also contains dark pools of lava or impact melt, as well as a fine example of a concentric crater. Its rilles are both radial and concentric, suggesting that they were formed by uplift of the floor following intrusion of subsurface magma. This would appear to be the explanation for most FFCs, which are generally (but not always) located near the edges of maria.

Although most rilles are to be found on or near mare terrain, there are occasional instances where they occur in highland areas. Perhaps the finest example is the rille system on the floor of the large, ancient crater Janssen in the southern highlands. Janssen does not appear to have been associated with volcanism, magma intrusion or tectonic stresses (although there are small floor

Humboldt and
its floor fractures.

Crater Humboldt,
imaged from Apollo 15.

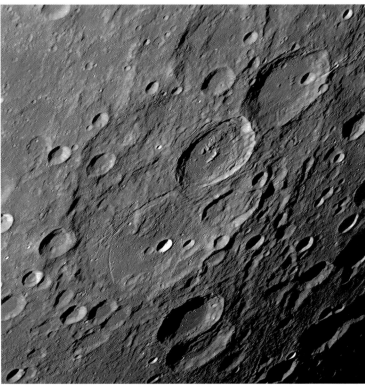

Rilles in Janssen.

fractures in the crater Fabricius that intrudes upon Janssen's northern flank), so it is rather difficult to account for its rilles.

The observer looking for a worthwhile observing programme might consider looking out for further examples of highland rilles. Indeed, the entire network of the Moon's rilles is still not fully charted, measured or understood, so there is useful work here yet to be done, both at the eyepiece and by using spacecraft imagery and datasets.

Faults

The observer should also look out for faults on the lunar surface. Like most rilles, these are tectonic features and they are caused by the slippage of land along crustal fractures. We have already seen examples of thrust faults in the form of wrinkle ridges on the mare surfaces, but there are also linear and strike/slip faults, where blocks of land are displaced either vertically or laterally. The best-known example of the former is the 'Straight Wall', or Rupes Recta, on the eastern edge of the Mare Nubium, where land to the west of the fault has slipped down about 400 m for a length of about 120 km (75 miles). Though it looks like a 'wall' telescopically, it is not in fact as cliff-like as it seems, with an average slope of only about 20°.

3D model of rille-fault near Bürg. A vertical exaggeration of ×5 has been applied to emphasize the transition.

The Ariadaeus Rille provides a particularly good example of lateral displacement caused by faulting. The horizontal forces that caused the graben to drop in the first place have also given rise to an offset about halfway along its course, and this is visible even in modest telescopes.

Given the association of both rilles and faults with similar tectonic forces, we should not be surprised to find examples of rilles turning into faults along their length. There is a particularly fine example of this on the gloomily named Lacus Mortis, southwest from the crater Bürg. A telescope of moderate aperture will suggest

The 'Straight Wall', also known as *Rupes Recta*.

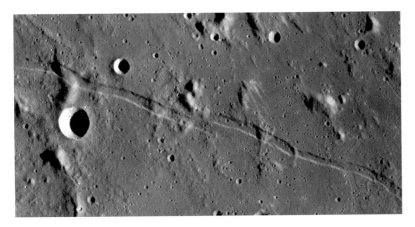

Lateral displacement along the Ariadaeus Rille.

the transition, but three-dimensional modelling using data from the LRO shows it to perfection.

Swirls and Ray Systems

Most of the landforms we have considered so far in this chapter are best observed under low angles of solar illumination, when shadows throw them into greatest relief. However, we might conclude this brief review by considering features that are best seen when the Moon is near full. We mentioned in Chapter Four the mysterious high-albedo swirls, such as Reiner Gamma, that appear to be associated with localized magnetic fields. Reiner Gamma itself is easily visible even in a modest telescope as the Moon approaches full, but only a few other examples are known, and it would make an interesting practical project to seek them out, again using both the telescope and spacecraft imagery. A basic list of swirls may be found online, but I am sure there are many more awaiting discovery.[7]

Superficially similar to swirls, in that they are extensive high-albedo surface markings, are the bright ray systems usually associated with relatively fresh impact craters. In reality, these are quite different from swirls in both nature and origin. The finest example is the system radiating from the crater Tycho, and at and around Full Moon it is clear that these rays extend for many hundreds of miles, draped over all kinds of terrain. The fact that they are bright and cover other features suggests that they are younger than the terrain they traverse. Their association with younger impact craters such as Tycho, Copernicus and Kepler, to give just three examples, implies that the rays, too, are a consequence of the impact process – and this turns out to be the case. They are ejecta deposits thrown out in a radial pattern by the violence of impact. Much of this ejected material created secondary cratering where it landed, excavating fresh (and therefore bright) materials that have not yet had time to darken as a result of 'space weathering'. You can learn much about relative ages by mapping the main ray systems visible in your telescope and noting the older terrain they cover and any instances of the rays themselves being interrupted.

Many ray systems form regular radial patterns around their crater at the impact site, but there are exceptions and from these we can learn much of interest. Take a look at the bright ray crater Proclus just to the west of the Mare Crisium. Note how the ray ejecta is arranged eccentrically around the crater, with most of it dispersed to the north, east and south of Proclus and with a clearly defined 'zone of avoidance' to the west.

Such a pattern is indicative of the fact that Proclus was formed by an impactor coming in at a very low angle from the west. We saw in Chapter Three that most impacts produce a point-source explosion giving rise to circular craters, more or less irrespective of the angle of impact at arrival (although lower-angle impacts can produce elongated craters). The real exception is when the angle of impact is very low, less than about 15°. Under those conditions ejecta is thrown out downrange, with a zone of avoidance uprange from the point of impact. Further rays are sent out sideways to form a butterfly-wing pattern, clearly seen around Proclus. Using such indicators it is instructive to search out further instances of oblique impacts. One particularly interesting example is provided

Messier and Messier A.

by the 'twin' craters Messier and Messier A on the Mare Fecunditatis. Here we see again the telltale signs of downrange 'comet-tail' and sideways 'butterfly' ejecta patterns, coupled with a clear zone of avoidance to the east indicating the direction from which the impactor arrived.

However, what is truly unusual about this pair is that the angle of impact appears to have been so low

Messier and Messier A from Apollo 11.

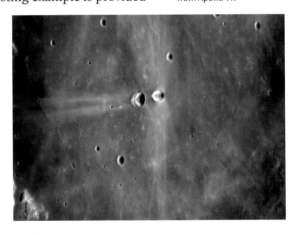

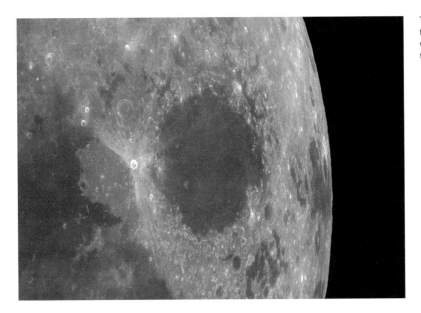

The Proclus rays, showing their eccentric disposition with a zone of avoidance towards the west.

as to be almost glancing. Indeed, it has been suggested that the impactor, having excavated Messier as a crater clearly elongated in an east–west direction, then ricocheted downrange to form Messier A, with its multiple west wall. This explanation is by no means certain, but the strange saddle shape of Messier when seen from spacecraft suggests that something extraordinary may have happened when these two craters were formed.

Eclipses

This chapter has so far tried to provide a review, albeit brief and by no means complete, of how the telescopic observer might start to explore at first hand the intriguing landforms visible on the Moon's surface. In the next section we shall move on to discover how the amateur might supplement his or her telescopic explorations by using data from spacecraft. But before that we must conclude this section by considering the opportunities and dangers posed by observation of eclipses. These provide spectacular, if relatively rare, observational opportunities, and the amateur will understandably want to make the most of them. Details of upcoming eclipses are widely advertised in astronomical magazines, as well as in specialist almanacs such as the *Handbook* of the British Astronomical Association. Eclipses of the Moon allow the observer to follow the progress of the Earth's shadow as it covers the disc of the Full Moon, and it is a worthwhile project to time the moment when the shadow reaches prominent craters and other landmarks. Lunar eclipses also vary significantly in colour and darkness: some are so dark that the Moon is hardly visible at all when covered by the Earth's shadow; on other occasions the eclipsed Moon remains easily visible as it turns a ruddy coppery colour. It all depends on the state of the Earth's atmosphere at the time of eclipse; when it is polluted by, for example, ash from recent volcanic eruptions, a lunar eclipse can be particularly dark. The Danjon scale, proposed by the French astronomer André-Louis Danjon in 1921, is used to indicate the relative darkness of lunar eclipses. It criterion-references eclipses on a scale of 0 (very dark, Moon almost invisible) to 4 (very bright) and, although it is inevitably subjective, it does provide a useful indicator.[8]

Observation of lunar eclipses poses no danger to the observer, but the same cannot be said of eclipses of the Sun. These, when total, are perhaps the most spectacular phenomena in the whole of nature. They are also relatively infrequent, so they draw large numbers of enthusiasts. But it is important to bear in mind that direct observation of the Sun carries real dangers and it can blind the unwary. The safest rule of thumb is *never look directly at the Sun without a proper solar filter*, and this rule applies whether or not you are using optical aids such as binoculars or telescopes. Even naked-eye observation of the unfiltered Sun can damage your eyesight permanently; unfiltered observation using binoculars or a telescope will almost certainly blind you for life. The danger is most acute during eclipses since as the Sun's disc is reduced by the encroaching

Moon, the temptation to 'take a quick peek' grows. Only when the Sun's disc is completely covered by the Moon is it safe to look directly, but even here you must anticipate the suddenness with which it will reappear. Observation of the partial phases of a solar eclipse must be carried out using reliable eclipse glasses for unaided observation and a properly fitting, full-aperture solar filter for a telescope or binoculars. So-called sun filters that fit over the eyepiece should be discarded: they will certainly crack once the heat of the Sun is focused on them, leaving the eye unprotected. Solar eclipses are unmissable, and everyone should witness the spectacle at least once in their lives – but not at the expense of loss of sight.

Observing the Moon using Spacecraft Datasets

Since the Ranger, Lunar Orbiter and Apollo missions of the 1960s a vast amount of data has been returned from generations of spacecraft that have visited the Moon's vicinity. Indeed, it is probably true to say that the amount of such data far outstrips the ability of the professional community to analyse it. Much of this information is freely available to all via the Internet, and it ranges from simple close-up imagery to specialized datasets addressing specific scientific questions. There is a huge opportunity here for the amateur lunar enthusiast, who can use such data to clarify and extend the knowledge gained from telescopic observation. It also allows the amateur armed with some knowledge of lunar geology to undertake investigations of real scientific value.

Considerations of space preclude here any attempt at a full discussion of the vast array of resources available, but the best places to start are with the data returned from the Lunar Orbiter (LO), Apollo and Lunar Reconnaissance Orbiter (LRO) missions. The five Lunar Orbiter missions, launched between 1966 and 1967, represented the first attempt at systematic high-resolution photography of the lunar surface. LOs I, II, III and V were primarily

The QuickMap home screen.

concerned with gathering data about potential landing sites for the Apollo programme, but LO IV provided a much wider survey of nearly the entire lunar surface. Photographic techniques have improved out of all recognition in the decades since, but the LO image gallery is still of great value. It can be accessed via the Lunar and Planetary Institute (LPI) website at www.lpi.usra.edu, from where images may be freely downloaded at various resolutions. Similarly, high-quality photographs from the various Apollo manned missions, including photographs taken from orbit, may be downloaded from the LPI Apollo image atlas at www.lpi.usra.edu/resources/apollo.

QuickMap colour relief overlay, showing the small crater Rhaeticus A.

But the jewel in the crown of lunar datasets is the information
returned by the Lunar Reconnaissance Orbiter (LRO) mission.
Launched in 2009, it is still operational at the time of writing. It has
returned a vast amount of scientific data including many thousands
of images showing the lunar surface at various resolutions, from
medium-resolution images of broad surface features down to
close-ups allowing the detection of individual boulders as well as
the hardware left behind by the Apollo lander missions. The LRO
website may be found at http://lroc.sese.asu.edu and it provides
a link via the 'Images' button at the top of the homepage to the
QuickMap toolset. This is a treasure trove for the lunar explorer,
and it has been used extensively to generate information and
imagery for the present volume. The QuickMap homepage
features an image of the Moon showing the current phase,
but you can deselect the phase view by unchecking the 'sunlit
region' box under the 'layers' menu in the top left-hand corner.

You can also select your preferred map projection (for
example, orthographic near-side or far-side, north or south
polar, or cylindrical) by clicking on the round icon in the top
right-hand corner of the 'layers' menu. Whichever projection
you choose, this then serves as a base map from which you can
zoom in on any feature by scrolling your mouse wheel, double-
clicking on a selected region or using the plus and minus buttons
in the extreme top left-hand corner of the page. Once you have
zoomed in on your area of interest, the 'layers' menu offers you
various overlay options, including a latitude/longitude grid,
nomenclature and the choice of normal or 'large' shadows.
The latter aid recognition of relief features. You can also overlay
useful datasets, including a colour-shaded relief map as well as
the Clementine colour-ratio/mineralogy data and GRAIL Bouguer
gravity gradients used earlier in this volume. There are many
more useful options under the 'layers' menu, and I urge you
to experiment.

Other valuable tools may be pressed into service by clicking on the spanner icon in the top right-hand corner. For example, clicking on the middle button revealed by the spanner (the 'line tool') allows you to use the mouse to draw a line between two points. Simply click once where you wish the line to start, drag across to your chosen finish point and then double-click. QuickMap will then magically generate a relief profile of the lunar surface along the line you have drawn, using data derived from the Lunar Orbiter Laser Altimeter (LOLA) results. This is an incredibly useful tool for understanding local topography, although do bear in mind that the graphic profiles generated are exaggerated in elevation (although true elevations in metres are given on the left of the profile).

The button to the right of the line tool (the 'search tool') is great fun. It allows you to draw a box around your chosen feature of interest, simply by clicking and dragging your mouse. If you then click on the highlighted box, followed by 'Query' and '3D live', a window will eventually appear entitled 'PIPE: Region of interest'. Scroll down this window until you reach a link marked 'NEW! Click here to access an experimental 3D visualization tool'. Clicking on that will reveal a three-dimensional model of the area selected, which you can tilt and rotate at will using your mouse. Again, this can reveal much topographical information about an area that you may previously have examined through your telescope.

What would our lunar observers of the past have given to have had access to such riches? Many of the topographical problems that exercised them can now be solved at the click of a mouse, even when the sky is cloudy. We have here only scratched the surface of what QuickMap can offer the lunar explorer. The site is constantly being updated to accommodate new developments and tools, and the instructions given will no doubt have to evolve in order to reflect those changes.

Conclusion

We have seen in the course of this volume how throughout
the history of observation of our Moon *individual* observers,
cartographers, geologists, theorists, as well as spacecraft, have
each woven their own unique strand into the tapestry of lunar
science. However, that science as a whole has taken shape as the
result of a *collective* endeavour, as each individual absorbed the work
of others and each generation learned from the ones that preceded
it. The same is true of lunar observation today. The amateur who
wishes to maximize his or her understanding and contribution
must both learn from the past and seek out the company of fellow
enthusiasts. This can best be done by joining an astronomical
society, either a local one or a national body such as the British
Astronomical Association (BAA) or the Society for Popular Astronomy
(SPA). The BAA, in particular, has an unmatched record in the
history of lunar study. Its Lunar Section kept the art of lunar
observation alive through the work of its amateur observers during
decades when professionals showed no interest in our satellite.
Today, too, it continues to support and organize the work of
amateur observers armed with modest equipment, and this is
of vital importance not just for the future of observational science,
but for the very fabric of our society.

We have heard much in recent years of the phenomenon
of 'citizen science' – the participation of ordinary people in the
collective scientific endeavour. It is an important phenomenon and
is likely to become even more so as scientific knowledge develops
in the future. It is now over half a century since the novelist-cum-
scientist C. P. Snow delivered his well-known Rede Lecture on 'The
Two Cultures'. Much of Snow's analysis is no longer taken seriously,
but in his lecture he lamented the gulf that had arisen between
practitioners of science and those who studied the humanities,
between 'scientists' and those he termed 'literary intellectuals'.

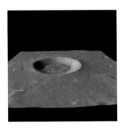

Generating a 3D model
of Rhaeticus A.

Deriving a graphic relief
profile of Rhaeticus A from
LOLA data.

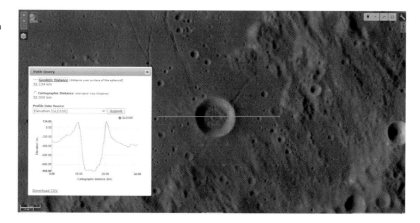

Separated by an abyss of mutual incomprehension, each side
spoke a language that was understandable only to itself. Literary
intellectuals, in particular, increasingly failed to understand the
advances made by a scientific world becoming more specialized
by the day. For Snow this was an acute problem even in 1959, when
his lecture was delivered. But it is an even more dangerous problem
today. For what kind of democratic society can we have if the majority
of the population does not even understand, let alone participate in,
the science and technology upon which that society is ever more
fundamentally based? Citizen science not only provides invaluable
and low-cost support for professional science, it draws large numbers
of ordinary people into a situation where they can understand and
evaluate scientific priorities and objectives.

Of course, some sciences lend themselves to this more readily
than others. It is difficult for the untrained amateur to make a
contribution in fields such as chemistry or nuclear physics, but
astronomy is different. As an observation-based science, it is more
amenable to non-specialized participation. The amateur observer
has the ability and opportunity to gather raw observational data on
a nightly basis and to share that data with amateur and professional
colleagues. As a result, there are many possibilities for fruitful pro-am
collaboration in the field of astronomy.

If this volume has helped to alert even one reader to those possibilities, and to the enrichment – personal and scientific – offered by careful study of our Moon in particular, then it has not been written in vain.

REFERENCES

1 OUR COMPANION MOON: FROM MIRROR TO MUSEUM

1 Philip J. Stooke, 'Mappaemundi and the Mirror in the Moon', *Cartographica: The International Journal for Geographic Information and Geovisualization*, XXIX/2 (1992), pp. 20–30.

2 For a fuller account of the Moon's significance in human culture see, for example, Edgar Williams, *Moon: Nature and Culture* (London, 2014).

3 Such a task has been capably carried out elsewhere. See, for example, Thomas L. Heath, *Greek Astronomy* (Cambridge, 2013).

4 For a useful tabulation of lunar statistical data see Patrick Moore and Robin Rees, *Patrick Moore's Data Book of Astronomy* (Cambridge, 2011), p. 26.

5 For an accessible fuller treatment of the Moon's importance see Joseph L. Spradley, 'Ten Lunar Legacies: Importance of the Moon for Life on Earth', *Perspectives on Science and Christian Faith*, LXII/4 (2010), pp. 267–75.

2 THE MOON AS A WORLD: OBSERVATION AND DISCOVERY IN THE TELESCOPIC AGE

1 Henry C. King, *The History of the Telescope* (London, 1955), pp. 30–31.

2 David Whitehouse, *The Moon: A Biography* (London, 2001), pp. 79–80.

3 Galileo Galilei, *Sidereus nuncius*, online edition based on the translation by Edward Stafford Carlos (London, 1880), available at http://homepages.wmich.edu/~mcgrew/Siderius.pdf, accessed 9 May 2017.

4 Allan Chapman, *Stargazers: Copernicus, Galileo, the Telescope and the Church* (Oxford, 2014), p. 141.

5 Galileo, *Siderius nuncius*, online edn.

6 Chapman, *Stargazers*, p. 141.

7 J. L. Heilbron, *Galileo* (Oxford, 2010), p. 151.
8 William P. Sheehan and Thomas A. Dobbins, *Epic Moon: A History of Lunar Exploration in the Age of the Telescope* (Richmond, VA, 2001), p. 6.
9 Heilbron, *Galileo*, p. 151.
10 See, for example, Sheehan and Dobbins, *Epic Moon*, p. 6, and Whitehouse, *The Moon*, p. 86.
11 Ewen A. Whitaker, *Mapping and Naming the Moon: A History of Lunar Cartography and Nomenclature* (Cambridge, 1999), p. 38.
12 Whitehouse, *The Moon*, p. 98.
13 Quoted in Whitaker, *Mapping and Naming the Moon*, p. 40.
14 Patrick Moore and Robin Rees, *Patrick Moore's Data Book of Astronomy* (Cambridge, 2011), p. 26.
15 Whitaker, *Mapping and Naming the Moon*, p. 60.
16 Ibid., p. 65.
17 Ivano Dal Prete, 'Cassini and Schröter's Valley', *The Moon: Notes and Records of the Lunar Section of the British Astronomical Association*, I/I (2011), pp. 9–12.
18 According to Jérôme de Lalande in his foreword to the 1800 revision of Bernard le Bovier de Fontenelle's *Conversations on the Plurality of Worlds*, trans. Elizabeth Gunning (Richmond, 2008), p. 20.
19 Ibid., p. 12.
20 Ibid., p. 70.
21 Ibid., pp. 54–5.
22 Joseph Ashbrook, *The Astronomical Scrapbook: Skywatchers, Pioneers, and Seekers in Astronomy* (Cambridge, 1984), pp. 236–44.
23 In J.L.E. Dreyer, ed., *The Scientific Papers of Sir William Herschel* (London, 1912), vol. I, p. 5.
24 Ibid., pp. 315–16.
25 See W. Sheehan and R. M. Baum, 'Observations and inference: Johann Hieronymous Schroeter, 1745–1816', *Journal of the British Astronomical Association*, CV/4 (1995), pp. 171–5.
26 See Ashbrook, *The Astronomical Scrapbook*, pp. 247–51.
27 Whitehouse, *The Moon*, p. 131.
28 Whitaker, *Mapping and Naming the Moon*, p. 131.
29 Ibid., pp. 133–5.
30 Ashbrook, *The Astronomical Scrapbook*, pp. 272–8.
31 See Sheehan and Dobbins, *Epic Moon*, pp. 197–202.
32 Ibid., p. 191.
33 For a fascinating study of this topic see K. Maria D. Lane, *Geographies of Mars: Seeing and Knowing the Red Planet* (Chicago, IL, and London, 2011).
34 Ibid., p. 25.

35 For an account of the Moon Hoax of 1835 see, for example, David S. Evans,
    'The Great Moon Hoax', *Sky and Telescope*, LXII: September 1981, pp. 196–8;
    October 1981, pp. 308–11.
36 W. H. Pickering, *The Moon: A Summary of the Recent Advances in Our Knowledge
    of Our Satellite, with a Complete Photographic Atlas* (London, 1904), p. 75.
37 Edmund Neison, *The Moon and the Condition and Configurations of its Surface*
    (London, 1876), pp. 111–30.
38 Patrick Moore and Peter Cattermole, *The Craters of the Moon: An Observational
    Approach* (London, 1967).
39 Thomas Gwyn Elger, *The Moon: A Full Description and Map of its Principal Features*
    (London, 1895).
40 Ibid., pp. 9–10.
41 Ibid., pp. 24–6.
42 Ibid., p. 22.
43 Walter Goodacre, *The Moon, with a Description of Its Surface Formations*
    (Bournemouth, 1931).
44 Walter Goodacre, 'Fauth's New Moon Charts', *Journal of the British Astronomical
    Association*, XLIII/5 (1933), p. 212.

3 THE MOON IN THE MODERN AGE

1 W. H. Pickering, *The Moon: A Summary of the Recent Advances in Our Knowledge
   of Our Satellite, with a Complete Photographic Atlas* (London, 1904), p. 89 onwards.
2 M. Loewy and P. Puiseux, *Atlas photographique de la lune* (Paris, 1896–1909).
3 Rudolf König, ed., *Joh. Nep. Kriegers Mond-Atlas* (Vienna, 1912).
4 *Photographic Lunar Atlas Based on the Photographs Taken at the Mount Wilson, Lick,
   Pic du Midi, McDonald and Yerkes Observatories*, compiled by G. P. Kuiper, D.W.G.
   Arthur, E. Moore, J. W. Tapscott and E. A. Whitaker (Chicago, IL, 1960).
5 M. A. Blagg and K. Muller, eds, *Named Lunar Formations*, 2 vols (London,
   1935).
6 Ewen Whitaker, 'The Founding of LPL: 1960–1972: The Early Days',
   www.lpl.arizona.edu, accessed 10 May 2017.
7 William K. Hartmann, 'The Founding of LPL: 1960–1972: Telescopes and
   Research: William Hartmann, on Photographic Lunar Research',
   www.lpl.arizona.edu, accessed 10 May 2017.
8 *Oxford English Dictionary*, www.oed.com.
9 William K. Hartmann, 'Discovery of Multi-ring Basins: Gestalt Perception in
   Planetary Science', in *Proceedings of the Conference on Multi-ring Basins: Formation
   and Evolution, Houston, Texas, 10–12 November 1980*, ed. P. H. Schultz and R. B.
   Merrill (New York, 1981), p. 79.
10 Ibid., p. 84.

11 David Whitehouse, *The Moon: A Biography* (London, 2001), p. 142.

12 Don E. Wilhelms, *To a Rocky Moon: A Geologist's History of Lunar Exploration* (Tucson, AZ, and London, 1993), p. x.

13 Grove Karl Gilbert, 'The Moon's Face: A Study of the Origin of Its Features', *Bulletin of the Philosophical Society of Washington*, XII (1893), p. 253.

14 Ibid., pp. 275–6.

15 Hartmann, 'Discovery of Multi-ring Basins', p. 82; Wilhelms, *To a Rocky Moon*, p. 14.

16 Wilhelms, *To a Rocky Moon*, p. 14.

17 Ralph B. Baldwin, *The Face of the Moon* (Chicago, IL, 1949), p. 61.

18 Ibid., p. 153, italics added.

19 Ibid., p. 65.

20 A. C. Gifford, 'The Origin of the Surface Features of the Moon', *Journal of the Royal Astronomical Society of Canada*, XXV (1931), pp. 74–8; based upon a paper first published in 1930.

21 'Report of the Meeting of the Association held on Wednesday, May 26, 1915', *Journal of the British Astronomical Association*, XXV/7 (1915), p. 326.

22 See Baldwin, *The Face of the Moon*, pp. 131, 135. The full argument is deployed throughout chapters 6 and 7.

23 Gifford, 'The Origin of the Surface Features of the Moon', p. 80.

24 Charles A. Wood, *The Modern Moon: A Personal View* (Cambridge, MA, 2003), p. 14. Wood's detailed but straightforward account of impact crater formation is one of the best I have come across, and I am indebted to it for my own summary here.

25 Baldwin, *The Face of the Moon*, p. 200.

26 Paul D. Spudis, *The Geology of Multi-ring Impact Basins: The Moon and other Planets* (Cambridge, 1993), pp. 30–33.

27 For a fuller discussion of ring formation than is possible here see Spudis, *The Geology of Multi-ring Impact Basins*, pp. 180–88.

28 Patrick Moore, Letter to Leslie F. Ball, 1 December 1949, archive of the Lunar Section, British Astronomical Association, London.

4 THE MOON IN THE AGE OF SPACECRAFT EXPLORATION

1 See, for example, Philip J. Stooke, *The International Atlas of Lunar Exploration* (Cambridge, 2007).

2 Arlin Crotts, *The New Moon: Water, Exploration, and Future Habitation* (Cambridge, 2014), p. 211.

3 See Richard M. Baum, 'Before Lunik: Imagination and the Other Side of the Moon', *Journal of the British Astronomical Association*, CXXV/1 (2015), pp. 23–37.

4 See, for example, Charles J. Byrne, *The Moon's Largest Craters and Basins* (Heidelberg, 2016), pp. 221–6.

5 Paul D. Spudis, 'Our Two-faced Moon', *Sky and Telescope*, CXXXI/4 (April 2016), p. 19.

6 Ralph B. Baldwin, *The Face of the Moon* (Chicago, IL, 1949), p. 23.

7 See Don E. Wilhelms, *To a Rocky Moon: A Geologist's History of Lunar Exploration* (Tucson, AZ, and London, 1993), p. 367. This table summarizes a fuller treatment in Wilhelms's *The Geologic History of the Moon*, USGS professional paper 1348 (Washington, DC, 1987), available online at http://pubs.usgs.gov, accessed 11 May 2017.

8 See Crotts, *The New Moon*, pp. 228, 251.

9 Ibid., p. 241.

10 Ibid., pp. 265–9.

11 See Brian Cudnik, *Lunar Meteoroid Impacts and How to Observe Them* (New York and London, 2009).

12 Crotts, *The New Moon*, ch. 9. See also Arlin Crotts and Cameron Hummels, 'Lunar Outgassing, Transient Phenomena and the Return to the Moon, II: Predictions and Tests for Outgassing/Regolith Interaction', *Astrophysical Journal*, DCCVII (2009), pp. 1506–23.

5 OBSERVING THE MOON

1 Antonín Rükl, *Atlas of the Moon*, revised updated edn, ed. Gary Seronik (Cambridge, MA, 2004).

2 Charles A. Wood and Maurice J. S. Collins, *21st Century Atlas of the Moon* (Wheeling, WV, 2012).

3 See, for example, R. Handy et al., *Sketching the Moon: An Astronomical Artist's Guide* (Heidelberg, 2012); Damian Peach, 'High Resolution Lunar and Planetary Imaging', in *Lessons from the Masters: Current Concepts in Astronomical Image Processing*, ed. R. Gendler (Heidelberg, 2013), pp. 233–59; Thierry Legault, *Astrophotography* (Santa Barbara, CA, 2014); and Nicolas Dupont-Bloch, *Shoot the Moon: A Complete Guide to Lunar Imaging* (Cambridge, 2016).

4 Harold Hill, *A Portfolio of Lunar Drawings* (Cambridge, 1991).

5 See Charles A. Wood, *The Modern Moon: A Personal View* (Cambridge, MA, 2003), p. 148.

6 A list of known concentric craters is given online at http://the-moon.wikispaces.com/Concentric+Crater.

7 For a list of swirls, see http://the-moon.wikispaces.com/swirl.

8 Details of the Danjon scale may be found on the NASA Eclipse website at http://eclipse.gsfc.nasa.gov/OH/Danjon.html.

FURTHER READING AND RESOURCES

This list does not contain all works referenced in the text. Nor does it pretend to be an exhaustive guide to further study. It merely offers a few resources that will help the reader to take the next steps in learning about our Moon.

BOOKS AND ATLASES

Baldwin, Ralph B., *The Face of the Moon* (Chicago, IL, 1949)

Cook, Anthony Charles, *The Hatfield Lunar Atlas: A Digitally Remastered Edition* (New York, Heidelberg, Dordrecht and London, 2012)

Crotts, Arlin, *The New Moon: Water, Exploration, and Future Habitation* (Cambridge and New York, 2014)

Dupont-Bloch, Nicolas, *Shoot the Moon: A Complete Guide to Lunar Imaging* (Cambridge and New York, 2016)

Grego, Peter, *The Moon and How to Observe It* (London, 2005)

—, *Philip's Moon Observer's Guide* (London, 2003)

Hill, Harold, *A Portfolio of Lunar Drawings* (Cambridge, 1991)

Moore, John, *Craters of the Near Side Moon* (NP, 2014 – available via Amazon)

—, *Features of the Near Side Moon* (NP, 2014 – available via Amazon)

Moore, Patrick, *Patrick Moore on the Moon* (London, 2001)

North, Gerald, *Observing the Moon: The Modern Astronomer's Guide* (Cambridge, 2007)

Rükl, Antonín, *Atlas of the Moon* (Cambridge, MA, 2004)

Sheehan, William P., and Thomas A. Dobbins, *Epic Moon: A History of Lunar Exploration in the Age of the Telescope* (Richmond, VA, 2001)

Stooke, Philip J., *The International Atlas of Lunar Exploration* (Cambridge, 2007)

Whitaker, Ewen A., *Mapping and Naming the Moon: A History of Lunar Cartography and Nomenclature* (Cambridge, 1999)

Whitehouse, David, *The Moon: A Biography* (London, 2001)

Wilhelms, Don E., *To a Rocky Moon: A Geologist's History of Lunar Exploration* (Tucson, AZ, and London, 1993)

Wood, Charles A., *The Modern Moon: A Personal View* (Cambridge, MA, 2003)
—, and Maurice J. S. Collins, *21st Century Atlas of the Moon* (Wheeling, WV, 2012)

MAPS

Map of the Moon (available from the British Astronomical Association at
     www.britastro.org/shop)
Murray, John, *Philip's Moon Map* (London, 2015)
Sky and Telescope's Mirror-image Field Map of the Moon (Cambridge, MA, 2007)
Sky and Telescope's Moon Map (Cambridge, MA, 2007)

ONLINE RESOURCES

Lunar Reconnaissance Orbiter Camera: http://wms.lroc.asu.edu/lroc
Lunar Reconnaissance Orbiter Camera – Farside: http://lroc.sese.asu.edu/
     posts/298
Lunar Terminator Visualization Tool: http://ltvt.wikispaces.com/LTVT
NASA Scientific Visualization Studio: Moon Phase and Libration: https://svs.gsfc.
     nasa.gov/4404
QuickMap: http://target.lroc.asu.edu/q3
Virtual Moon Atlas: https://sourceforge.net/projects/virtualmoon

ACKNOWLEDGEMENTS

I have been helped in the course of my work on this project by advice received from Richard Baum, Barry Fitz-Gerald and Peter Morris. I am extremely grateful to them for taking the trouble to read rough drafts and for suggestions that have improved both the content and the approach of this book. I would also like to thank those who have provided illustrative material and whose contributions are acknowledged elsewhere.

Photo Acknowledgements

The author and publishers wish to express their thanks to the below sources of illustrative material and/or permission to reproduce it:

Photos courtesy Paul Abel/British Astronomical Association: p. 128 (all); Leo Aerts: p. 146; photos author: pp. 36, 80, 111, 113 (bottom), 126, 138 (left and right), 148, 156; from Wilhelm Beer and Johann Madler, *Der Mond nach seinen kosmischen und individuellen Verhaltnissen* (Berlin, 1837): p. 47; photos courtesy British Astronomical Association: pp. 49, 53 (left and right), 58, 60, 135, 137, 150; © Mike Brown: p. 147; from Gilbert Fielder, *Lunar Geology* (London, 1965)/reproduced courtesy of Gilbert Fielder: p. 92; © Chris Hooker: p. 158; © Simon Kidd/NASA: p. 89 (left and right); from Johann Krieger, *Mond-Atlas* (Vienna, 1912): p. 65; from Michel van Langrenus, *Plenilunii Lumina Austriaca Philippica* (Paris, 1645)/image National and University Library of Strasbourg: p. 31; from Wilhelm Gotthelf Lohrmann, *Topographie der Mondoberfläche* (Leipzig, 1824): p. 45; photos LRO/NASA: pp. 48, 114 (left and right); from Patrick Moore, *Survey of the Moon* (London, 1963)/ reproduced courtesy of the Trustees of the Sir Patrick Moore Heritage Trust: pp. 86, 90; from Patrick Moore and Peter Cattermole, *The Craters of the Moon* (London, 1967)/ reproduced courtesy of Peter Cattermole and the Trustees of the Sir Patrick Moore Heritage Trust: p. 87; photos NASA: pp. 15, 18, 19, 69, 72, 78 (left), 81, 84, 88 (top), 91, 96, 98 (left and right), 102, 104 (left and right), 112, 119, 122, 123, 125, 144, 145, 149, 152 (top), 154 (top), 157, 161 (top and bottom), 164, 165; photo NASA (Apollo Archives): p. 91; photos NASA/QuickMap: pp. 113 (top), 140 (all), 153, 154 (bottom); photo NASA Scientific Visualization Studio: p. 83; from James Nasmyth and James Carpenter, *The Moon Considered as a Planet, a World, and a Satellite* (London, 1874): p. 56; from *New York Sun* (1835) p. 54; © Damian Peach: pp. 33, 50, 77, 79 (left), 103, 108, 109, 110 (left and right), 136, 139, 151, 152 (bottom); © Damian Peach/NASA: p. 88 (bottom left and right); photos Science Museum Pictorial/Science and Society Picture Library: pp. 39, 41, 42, 63; photo courtesy William Sheehan: p. 35; © Alan Tough: p. 142; Universal Images Group Ltd/

Science and Society Picture Library: p. 34; photo courtesy Ewen Whitaker and William Sheehan: p. 26; photos Don Wilhelms (courtesy of Universal Images Group Ltd/ Science and Society Picture Library): p. 97 (left and right); diagrams courtesy Charles A. Wood/*Sky & Telescope* (an imprint of F+W Media Inc.): pp. 78 (right), 79 (right); photo Worldspec/NASA/Alamy: p. 6.

INDEX

Page numbers in **bold italics** refer to illustrations